AF661365

Mulberry for Animal Production

Mulberry for Animal Production

by

R.K. Patnaik

BIOTECH

2008

Biotech Books

Delhi - 110 035

2008 © Publisher

ISBN10 81-7622-187-2
ISBN13 978-81-7622-187-0
EAN: 9788176221870

Information contained in this work has been published by Biotech Books and has been obtained by its author(s)/editor(s) from sources believed to be reliable and are correct to the best of their knowledge. However, the publisher and its author(s) make no representation of warranties with respect of accuracy or completeness of the contents of this book, and shall in no event be liable for any errors, omissions or damages arising out of use of this information and specifically disclaim any implied warranties or merchantability or fitness for any particular purpose.

All rights reserved. Including the right to translate or to reproduce this book or parts thereof except for brief quotations in critical reviews.

Published by : **BIOTECH BOOKS**
1123/74, Tri Nagar,
Delhi - 110 035
Phone: 27383999
e-mail: biotechbooks@yahoo.co.in

Showroom : 4762-63/23, Ansari Road, Darya Ganj,
New Delhi - 110 002
Phone: 23245578, 23244987

Printed at : **Chawla Offset Printers**
New Delhi - 110 052

PRINTED IN INDIA

Preface

Mulberry is present all over the world and its unique characteristics of yield, palatability and nutritive value make it a valuable resource for improving and intensifying a variety of livestock production practices. It was one of the first domesticated forages in the world and has been the subject of intensive research in various countries over the last few decades. This has resulted in higher yields of better quality foliage to be converted into silk.

Over several thousands of years, mulberry species and varieties have been selected and improved to feed the silkworm, which is nutritionally very demanding. The aim has been to produce greater quantities of leaves of higher quality under a wide range of conditions. Although, the principles of mulberry cultivation for sericulture purposes should apply to its cultivation as forage for feeding farm animals, there are certain important differences. Frequency of harvesting and planting density, can be intensified if the goal is to feed ruminants, since the quality is already high enough and the main purpose is to increase overall biomass yields. The issues of maintenance of soil fertility and plant persistence become important if huge quantities of nutrients are extracted from the soil in the biomass under cut-and-carry systems.

Since the quality of silk production is directly proportional to the quality of leaves used as the exclusive feed for these worms, leaf quality is of utmost importance in sericulture. Huge variation in the plant type foliar production, nutritive value and palatability of the leaves exists. From the commercial viewpoint, a shorter tree/ shrub with enhanced foliar production/vegetative growth and thicker and juicer leaves is highly desirable. Most of these traits are quantitative in nature and hence difficult to manipulate cumulatively by conventional approaches. The goals of mulberry improvement are thus directed towards development of high productivity genotypes with quality leaves.

This book provides an in-depth analysis of the production of healthy mulberry crop for harvesting better quality silk. It is designed for students and teachers of sericulture, scientists working in the field, sericulture extension workers and for general academic professional interest.

Editor

Contents

1

Introduction

Mulberry, originated from temperate zones in Asia, has now spread throughout the world. In tropical Central and South America, it has adapted excellently. It is a perennial tree or shrub that is easily propagated, with fast growth and vigorous shooting. It develops a strong vertical and profuse horizontal root system. These features improve physical soil conditions and allow better water conservation. As forage, mulberry has shown excellent organoleptic intake for livestock.

Protein content varies from 14 to 22 percent on a dry matter (DM) basis and *in vitro* digestibility between 70 and 80 percent. Under tropical conditions, where conditions favour plant growth, mulberry culture has been practised under various densities and cutting frequencies, which result in large yield differences. In Guatemala, Blanco obtained 19 tonnes of DM/ha in four cuts every nine weeks, with 30 cm between plants and a cutting height of 75 cm.

In Costa Rica, Benavides, Lachanx and Fuentes used densities of almost 23 000 plants per ha, with DM yields from 21 to 28 tonnes/ha. Rodríguez, Arias and Quiñones worked in Guatemala with planting distances of 60 and 80 cm between plants and harvesting frequencies of six, nine and 12 weeks, with DM yields of 1 to 4.6 tonnes/ ha/harvest. These results suggest that yield increases with density.

In Japan two planting densities are used, 10 000 and 20 000 plants/ha for the traditional and intensive production systems, respectively.

Forage Production

The table 1 presents DM yields by density, height and cutting frequency. Density strongly influenced yields. Yields dropped 37 percent from planting densities 60 to 90 cm and 13 percent from 90 to 120 cm. Leaf:stem proportion did not change among planting densities despite yield differences. More DM was produced at the 60 cm cutting height because of more leaves.

Table 1. Annual dry matter production (tonnes/ha) of mulberry depending on planting density, cutting height and frequency

	Cutting		Fraction			Leaf
Spacing (cm)	Height (cm)	Frequency (days)	Whole plant	Stems	Leaves	Stem ratio
60	30	56	18.3	7.1	11.2	1.60
60	30	84	25.1	11.4	13.7	1.20
60	30	112	40.6	21.6	19.0	0.88
60	60	56	24.4	9.2	15.3	1.75
60	60	84	35.8	16.3	19.5	1.19
60	60	112	30.9	15.8	15.1	0.94
90	30	56	10.2	4.0	6.2	1.63
90	30	84	16.5	7.8	8.6	1.11
90	30	112	26.8	14.5	12.3	0.85
90	60	56	11.2	4.5	7.3	1.64
90	60	84	21.5	9.3	12.2	1.32
90	60	112	20.8	10.3	10.4	1.04
120	30	56	10.0	4.2	5.9	1.40

Contd....

Contd.						
120	30	84	15.5	7.5	8.0	1.08
120	30	112	19.2	10.4	8.8	0.86
120	60	56	10.1	39.7	60.8	1.63
120	60	84	15.8	7.2	8.6	1.22
120	60	112	20.9	10.9	10.0	0.96
Means						
60			27.5	12.3	15.1	1.36
90			16.3	7.5	9.0	1.35
120			14.1	6.6	7.5	1.27
	30		18.4	8.7	9.8	1.26
	60		20.1	8.9	11.3	1.38
		56	14.0	5.5	8.7	1.61
		84	21.7	9.9	11.8	1.19
		112	26.5	13.9	12.6	0.93

Total and stem DM yields increased linearly with the length between harvests, but leaf yield only increased from 56 to 84 days. Leaf proportion decreased linearly with harvest length. Up to 100 days there were more leaves than stems. Leaf production is little affected by cutting frequency. The interaction planting density by cutting height was significant for the whole biomass and highly significant for leaf yield. Density by frequency interaction was not significant for whole plant, leaf and leaf:stem ratio, but highly significant for stem yield.

Height by frequency interaction was significant for all DM yields but not for leaf:stem ratio, despite the wide range observed. At 30 cm of cutting height DM increases with cutting length, but at 60 cm height, DM yields do not raise further 84 days. Similar effects are seen with leaf and stem yields. The three-way interaction - height, frequency and density - was significant for the three variables measured, but not for leaf:stem ratio.

In China, traditional density is 10,000 plants/ha and 25 000 for intensive cultivation. Lin, Yu and Hsieh, studied spacing effect on leaf production, and found similar results to those of the four times per year harvest of the present experiment. Higher shooting per plant does not compensate for higher shooting per hectare at higher densities. In general terms, larger spacing reduces light competition and high-density plantations respond to this competition. The long-term effects on the root system and leaf yield are unknown.

Cutting height had a little effect on yield, increasing by 1.7 tonnes DM/ha/year from 30 to 60 cm. Both of these heights are considered low for Chinese practices. This increment was due to the leaf fraction. In Turrialba annual plant yields were 2.32 kg cutting at 50 cm and 2.12 kg at 100 cm in low-density plantations. Some reports recommend low cutting heights for 30 000 plants/ha; medium height for 12 000 plants/ha and high cutting for densities below 6 000 plants/ha. Greater leaf production at 60 cm suggests that leaf and stem yields should be studied at various cutting heights under high light competition in dense plantations.

Frequency of defoliation had a marked effect on yield. Greater yields were observed at longer cutting frequencies. This is related to the regrowth delay period and to plant nutrient reserves. The results clearly show that annual production is lower at short intervals. Similar responses were obtained by Rodríguez, Arias and Quiñones when studying frequencies of harvesting of six, nine and 12 weeds in two different periods in Guatemala. In the second year the yield was triple but the frequency effect was maintained.

When studying plant growth post-harvest, it was observed that stem buds do not regrowth immediately. It takes four to ten days for the shoot to reach 1 cm. Afterwards, the re-growth continues with two to three buds on side branches. The density of re-growth seems proportional to the number of cuts. These observations indicate that plant structure and morphology, reflected on leaf and stem proportion, change with cutting frequency. After each harvest there was a little flux of sap in each cut. The results of this experiment show that plant stress, caused by cutting frequency has, in the short term, a negative effect on annual biomass yield. In Asian countries, often leaves are only picked every two to three months, and there is an annual pruning. In more intensive systems, leaf harvest/high pruning in the spring is combined with leaf harvest/low pruning in the winter.

Mulberry (*Morus* sp.) has made new advances worldwide as a forage plant because of its high protein content and high digestibility. However, its management is not yet defined, and there are various alternatives when is it destined to be used for domestic farm animals.

In Brazil, mulberry is known as feed for the silkworm (*Bombyx mori*) and it is cultivated at approximate spacing of 1 m^2/plant, with cuts close to the ground at 91 days-intervals. Management of mulberry as a shrub or tree has been encouraged by some experienced veterans, such as Giuseppe Briani, Marino Serpa and Mário Hashimoto, who are familiar with the practices utilised in Europe and Japan. These experts defend the advantages of the system: greater production and quality, better exploitation of soil, longevity and possibility of associated crops, in the limits of roads or property boundaries. Briani states that trees older than 200 years in Assisi produce about 15 kg of leaves per tree/year.

In China, investigations have shown that in Zhejiang and Jiangsu Provinces, when the number of branches and shoots is below 75 000/ha, the annual leaf yield is about 15 tonnes; between 90 000 and 150 000, the yield is 22.5-37.5 tonnes; but above 150 000, the yield will decrease.

The ideal number is between 100 000 and 120 000. High-yield mulberry plantations have a good colonial structure starting with a planting density of 12 000-15 000 plants/ha, the number of crowns at 37 500-45 000; the number of branches and shoots at 105 000-120 000 and the total length of the branches at 150 000-180 000 m per ha. Some high-yield plantations have a density at 22 500-30 000 trees/ha with four branches and shoots per tree, for a total of branches/shoots. Apart from their numbers, the length and diameter of branches and shoots are also directly related to leaf yield.

Varieties

When a new mulberry field is established, high-yield and superior quality varieties, adaptable to the local environment, should be used. Among the Husang varieties, Heyebai is a high-yield variety widely cultivated in Zhejiang and Jiangsu Provinces.

Fertilisers

There is a direct positive relationship between fertiliser dosage and quantity and quality of mulberry leaves. However, there is a reduction in the incremental response to fertiliser application as the dose increases. High-yield mulberry plantations require a plentiful supply of fertilisers and water. Timing and application method should be adequate to obtain maximum response.

Plant Shape and Age

Under similar conditions, low-trunk trees can be formed in a short period, and their leaf yields are higher than and medium-trunk trees, which take longer to develop. Harvest index and leaf yield of trees are low because of the high proportion of the non-assimilating organs. Since the 1970s, many high-yield plantations with low-trunk or trunkless mulberry trees have been established in China. The varieties planted have good characteristics, such as fast growth and high harvest index, but the highly productive life is shorter compared with high and medium-trunk trees.

Compared with mature trees, both young and old trees have a lower leaf yield. Young trees need to form structural organs, and old trees have lost physiological capacity. Therefore, attempts should be made to develop plants quickly and to lengthen the high-yield period as much as possible. Apart from the above factors, mulberry leaf yield is also related to methods and frequency of harvesting, as well as to crop management.

Distribution and Use of Mulberry

According to FAO, mulberry is both a temperate and a subtropical plant grown in many regions of the world, predominantly in eastern, southern and southeastern Asia, southern Europe, southern North America, northwestern South America and parts of Africa. In Africa, it has been reported to occur in humid, subhumid and semi-arid areas at an altitude up to more than 1 000 m above sea level. It is not well known when mulberry was introduced in Tanzania but it appears to have been in the country for many decades. In the humid areas of the northern highlands mulberry trees are incorporated into the intensive farming systems and are widely used to feed sheep and goats in a cut-and-carry system. In some situations mulberry is used as live fence. However, there have been reports that some missionaries have established mulberry trees for wine making, using the fruit.

Apart from the highlands of the north, mulberry is also present in the semi-arid areas of central Tanzania, in the coastal belt and in the southern highlands. So far, there have been no formal studies on the distribution and use of mulberry in the country. However, survey studies in some villages in two districts of the semi-arid areas of central Tanzania showed that mulberry trees were found in 5 percent of homesteads. The trees were grown close to the dwellings and the number of trees per household did not exceed five. Some trees were found away from the homesteads, where they were left by farmers who previously occupied the land. The trees were used for a range of purposes: for fruit, shade, vegetables, medicinal purposes and as fuel wood. The leaves were rarely used for livestock feeding, since most farmers were not aware of this use.

Productivity

There are several natural and management factors affecting the productivity of mulberry, such as soil and climatic conditions, mode of propagation and harvesting techniques.

Table 2. Mean yield of mulberry, by fractions, under different spacings at various periods of the year

Harvest (years)	*Spacing*	*Yield/plant (kg DM)*				*Yield/ha (tonnes/DM)*			
		Leaf	*Stem*	*Bark*	*Total*	*Leaf*	*Stem*	*Bark*	*Total*
1st-2[3]	S1	0.59	0.97	0.18	1.74	16.9	28.2	5.3	50.4
	S2	0.65	1.37	0.28	2.30	3.4	7.1	1.4	11.8
	S3	0.63	1.12	0.22	1.97	8.5	15.1	3.0	26.6
1st-2	S1	0.08	1.65	0.46	2.19	2.3	47.8	13.2	63.3
+ 190 days	S2	0.31	2.14	0.54	2.98	1.6	11.0	2.8	15.4
	S3	0.14	2.00	0.49	2.64	2.0	27.1	6.6	35.7
2nd-120 days	S1	0.14	0.12	0.05	0.31	4.0	3.5	1.4	9.00
	S2	0.13	0.14	0.06	0.33	0.7	0.7	0.3	1.7
	S3	0.11	0.14	0.06	0.30	1.5	1.9	0.8	4.1
2nd-190 days	S1	0.05[b]	0.16	0.02	0.22	1.3	4.5	0.6	6.4
	S2	0.06	0.17	0.02	0.25	0.3	0.9	0.1	1.3
	S3	0.05	0.17	0.02	0.23	0.6[c]	2.3	0.3	3.1

In general, longer intervals between defoliation have increased total yield; however, the proportion of inedible wood may also increase, leading to a decline in forage quality. On the other hand, as plant spacing is reduced, yield per plant decreases owing to competition, but total forage yield per unit area increases, as does the leaf:wood ratio. This was confirmed by Shayo in a trial to study biomass production at different spacing and harvesting intervals in the semi-arid areas of central Tanzania (Table 2).

Table 3. Proportion of mulberry fractions as affected by spacings and harvest time

Type	*Treatment*	*Part of the plant (%)*		
		Leaf	*Stem*	*Bark*
First harvest at 2 y	S1	33.6	55.9	10.5
	S2	28.3	59.6	12.2
	S3	31.8	56.8	11.4
	Mean	31.2	57.4	11.4
First harvest at 2 y + 190 days	S1	3.6	75.5	20.9
	S2	10.2	71.7	18.1
	S3	5.5	75.9	18.6
	Mean	6.4	74.4	19.2
Regrowth at 120 days				
	S1	44.7	39.4	15.9
	S2	39.3	42.5	18.2
	S3	36.1	45.4	18.4
	Mean	40.0	42.4	17.5
Regrowth at 190 days	S1	20.5	70.1	9.3
	S2	23.7	66.8	9.5
	S3	20.3	71.9	8.0
	Mean	21.5	69.6	8.9

Although yield per season increased with harvesting interval, the amounts and nutritive value of edible components decreased considerably due to senescence and drop of leaves. Tikader *et al.*, found that maximum mulberry leaf yields could be obtained by harvesting three times per year. The results in Table 2 show that mulberry harvested twice, at the end of rainy season and in the middle of the dry season, gave considerably higher amounts of fodder.

It is interesting to note that mulberry harvested during the peak of the dry season developed considerable numbers of shoots. As reported by Walker, regeneration of new shoots after pruning is independent of rainfall. The advantage with the new shoots is that they are more tender and nutritious than ordinary branches and grow more rapidly. This is also reflected by higher proportion of leaf:stem in younger plants and regrowths than older plants (Table 3).

Nutritive Value

Mulberry leaves are highly digestible and contain high concentrations of crude protein (CP) and minerals, and low cell wall contents (Table 4). This suggests that mulberry could be used to minimise nutrient deficiencies faced by grazing animals in the semi-arid areas of central Tanzania. Furthermore, the bark is reasonably digestible, with CP concentrations higher than in dry season pastures and crop residues. Older leaves contain lower concentrations of CP than young leaves (Table 4). Young leaves and bark are more degradable than old leaves and bark.

Table 4. Chemical composition and digestibility of mulberry leaves and bark in central Tanzania

Parameter (% DM)	Part of the plant			
	Leaf	*Bark*	*Leaf*	*Bark*
Ash 14.3	6.1	13.3	6.3	
Crude protein	18.6	7.8	14	8.7
Neutral detergent fibre	24.6	46.8	27.6	44.5
Acid detergent fibre	20.8	36.9	25.1	38.3
Hemicellulose	3.8	9.9	2.5	6.2
Acid detergent lignin	8.1	7.1	-	-
Cellulose	12.6	29.7	-	-
Acid detergent insoluble ash	2.5	1.1	-	-
In vitro DM digestibility	82	60	-	-
In vitro organic matter digestibility	-	-	89	85

The rate of in sacco degradation of leaves was high, 70-80 percent at 24 hours of incubation. The study also showed that degradability of the old bark was maximal after 72 hours. Another advantage of mulberry leaves is that they contain considerably lower levels of total phenolics compared with leaves and pods of most native browse trees and shrubs in central Tanzania, such as Faidherbia albida, Acacia tortilis, Acacia nilotica, Delonix elata and Dichrostachys cinerea. Phenolic compounds have a capacity to bind to carbohydrates, plant proteins, salivary mucoprotein and gastrointestinal enzymes, thereby reducing protein and cell wall digestion.

2

Mulberry: An Exceptional Forage

Mulberry belongs to the Moraceae family (subtype angiosperms; class dicotyledons; subclass urticales) and there are several species: *Morus alba, M. nigra, M. indica, M. laevigata, M. bombycis*, etc. which have been used directly, or through crossings and induced mutations, for the development of varieties to support silkworm production. The diploid *M. alba* (2n=2x=28) is the species most widely spread, but polyploid varieties, which originated in various research stations in Asia, show greater leaf yields and quality. In general, polyploid varieties have thicker and larger leaves of a darker green colour, and produce more leaves.

There is a large variation in leaf production and in leaf quality (e.g. protein content) among the many species and in the varieties and cultivars grown at different locations under a wide range of soil and environmental conditions, indicating the huge potential for identifying suitable germoplasm for most sites. Many of the references on mulberry in the literature do not specify which species or varieties were used. Names are often given based on leaf features. In many cases, locally grown varieties (native or criolla) seem to perform adequately, since they are probably well adapted to local conditions.

The main use of mulberry globally is as feed for the silkworm but, depending on the location, it is also appreciated for its fruit (consumed fresh, in juice or as preserves), as a delicious vegetable (young leaves and stems), for its medicinal properties in infusions (mulberry leaf tea), for landscaping and as animal feed. In Peru the multiple uses of mulberry have been recognised. There are several countries where mulberry is utilised traditionally as a feed in mixed forage diets for ruminants, such as in certain areas of India, China and Afghanistan.

In Italy there have been several studies on the use of mulberry for dairy cows and other domestic animals and in France a research project was undertaken to introduce mulberry in livestock production. But it was only in the 1980s that specific interest in the intensive cultivation and use of mulberry as animal feed started in Latin America. It is surprising that a plant that has been improved for leaf quality and yield to feed the silk worm, which has high nutritional feed requirements, has received such limited attention from livestock producers, technicians and researchers.

Like several significant breakthroughs in science and technology, the discovery of the value of mulberry as a high-quality feed in Latin America happened serendipitously. A Costa Rican farmer of Chinese origin, whose silk project had failed, fed mulberry leaves to his goats and was impressed by the palatability of the leaves and by the performance of his animals. He communicated his observations to scientists at the Tropical Agriculture Research and Training Center (CATIE), who were receptive to the farmer's news and forward thinking enough to include mulberry in their tree fodder evaluations and later in agronomic and animal performance trials. In Africa, the International Centre for Research in Agroforestry (ICRAF) in Kenya and the Livestock Production Research Institute in the United Republic of Tanzania have also conducted successful agronomic and animal trials by themselves, apparently without being aware of the interest elsewhere.

In Spain, Mulberry was recommended to be planted in association with other crops like maize, potatoes, vegetables, alfalfa and fruit trees, always controlling spacing and pruning to avoid light competition. With higher densities, yields of 30,000 kg/ha have been obtained. Yields are related to plantation age and more specifically with trunk diameter. Annual leaf production in monoculture increases from 6,500 kg to 33,500 kg/ha from the first to the seventh year. In good land, leaf yield per plant varies from 9 to 70 kg when average trunk diameter increases from 7 to 55 cm. With 22.5 tons of human faeces and 300kg of ammonium sulfate, fresh leaf production can reach 13 ton/ha/year.

In Paraguay, leaf yields of 20,000 kg/ha have been obtained in 4-year plantations harvesting at 30cm from the surface. Work with widely spaced plants in Turrialba a yield of 2.32 kg of dry matter per plant was calculated per year cutting at a height of 50 cm. Cutting at 1m, the yield decreased to 2.16kg. However, leaf production was 1kg for both cutting heights. With cuts every 60,120 and 180 days, total dry matter production was 1,64, 2,17 and 2,86 kg/plant/year, respectively. However, leaf production declined from 1.11 to 0.84kg between 60 and 180 days.

Chemical Composition of Mulberry

The results of chemical composition of mulberry fractions from various authors are given in Table 1. Crude protein content in leaves varies from as low as 15 percent to 28 percent depending on the variety, age of the leaves and growing conditions. In general, crude protein values can be considered similar to most legume forages. Fibre fractions are low in mulberry leaves compared with other foliages. Shayo reported lignin (acid detergent lignin) contents of 8.1 percent and 7.1 percent for leaves and bark, respectively.

Table 1. Chemical composition (percentaje of dry matter) of mulberry

Variety	*CP*	*CF*	*NDF*	*ADF*	*EE*	*Ash*	*Ca*	*P*
Leaf								
Hebba	15.9	12.6			7.1	15.9	2.42	0.24
Izatnagar	15.0	15.3			7.4	14.3	2.41	0.24
Palampur	15.0	11.8			5.1	15.5		
Parbhani	22.1	5.9			3.9	13.4	3.3	1.43
Kanva-2	16.7	11.3	32.3		3.0	17.3	1.80	0.14
Mpwapwa	18.6		24.6	20.8		14.3		
Dominican	20.0			23.1	4.0	4.5	2.70	
Criolla	19.8						1.90	0.28
Tigreada	21.1						2.74	0.38
Indonesia	20.1						2.87	0.33
Leaf & young stem								
Tigreada	27.6	13.2				10.4		0.20
Indonesia	24.3	15.3				11.2		0.29
Criolla	27.6	16.9				11.8		0.26
Acorazonada	25.2	14.1				13.4		0.15
Koruso 212	11.0	10.0	22.0	20.6	5.9	13.9	3.13	0.37
Koruso 213	8.0	11.8	24.7	24.5	5.3	19.3	4.76	0.37
Young stem								
Criolla	11.3						1.33	0.29
Tigreada	11.7						1.38	0.33
Indonesia	11.9						1.53	0.43
Dominican	4.7			48.2	1.7	1.3	1.61	

Stem								
Dominican	3.8			50.2	1.0	1.8	1.10	
Mallur	11.5	34.0			2.7	9.32	1.56	0.20
Bark								
Mpwapwa	7.8		46.8	36.9		6.1		
Whole plant								
Dominican	11.3			34.4	1.6	1.9	2.10	

A striking feature of mulberry leaves is the mineral content, with ash values up to 25 percent. Typical calcium contents are around 1.8 to 2.4 percent and phosphorus 0.14 to 0.24 percent. Espinoza, Benavides and Ferreire, found potassium values of 1.90-2.87 percent in leaves and 1.33-1.53 percent in young stems, and magnesium contents of 0.47-0.63 percent for leaves and 0.26-0.35 percent for young stems. Table 2 shows the digestibility of mulberry. As can be seen, leaf digestibilities in vivo (goats) and in vitro are very high (>80 percent) and total digestibility is equivalent to that of most tropical forages.

Table 2. Digestibility of mulberry

Method	Fraction	Digestibility (%)
In vivo (goats)	Leaf	78.4-80.8
In vitro	Leaf	89.2
	Leaf	80.2
	Leaf	89-95
	Stem	37-44
	Total	58-79
	Leaf	82.1
	Bark	60.3

The degradation characteristics of mulberry, determined by the nylon bag technique, are presented in Table 3.

Leaves would be completely degraded if they remained in the rumen for enough time.

Table 3. In sacco degradation of mulberry

Fraction	Parameter			
	a	b	a + b	c
Leaf	35.7	64.0	99.7	0.0621
Whole plant	30.4	46.2	76.6	0.0667
Leaf & young stem	27.8	48.95	76.8	0.0300

Table 4. Amino acids and N content of mulberry varieties and soybean meal

Compound	Soybean meal		Mulberry		
	content (mg/g DM)	%1	Content (mg/g DM)	SD	%1
Non essential amino acids	n.a		108.93		53.3
Essential amino acids (EAA):					
Lysine	32.92	6.7	12.33	2.58	6.0
Methionine	7.30	1.5	2.99	0.61	1.5
Threonine	20.34	4.1	10.52	1.75	5.2
Valine	26.29	5.3	12.83	2.17	6.3
Isoleucine	26.85	5.4	10.04	1.88	4.9
Leucine	39.55	8.0	19.45	3.10	3.1
Tyrosine	14.38	2.9	7.40	1.39	3.6
Phenylalanine	25.51	5.2	12.26	2.06	6.0
Histidine	12.92	2.6	4.61	0.82	2.3
Trytophane	6.97	1.4	na	-	-
Total EAA	213.03	43.1	92.43	-	45.3
Ammonia (NH_3)	na		2.89	0.54	1.4
Total (AA + NH_3)	494.38	100	204.25		100
Nitrogen (percent)	7.91		4.36	9.63	

The average amino acid composition and N content of 119 mulberry varieties grown experimentally in Japan are shown in Table 4. Tryptophane was not included in the analysis. As can be seen from the data, essential amino acids are over 46 percent of total amino acids. It can be calculated from the table that the average nitrogen (N) is 16.6 percent of the total molecular weight of the mulberry amino acids (plus ammonia), and thus the converting factor from N to mulberry protein is 6.02. The 204.3 mg of amino acids per g of protein is equivalent to 3.47 percent N, which is 80 percent of total N in mulberry leaves.

Once tryptophane is subtracted, the difference, a non-protein fraction, is likely to be composed of nucleic acids and other unidentified N compounds. The most important protein in mulberry leaves, as in most leaves, is ribulose-1.5-bisphosphate carboxylase (RuBisCO) whose active site is responsible for carbon fixation. Nitrogen in RuBisCO can be 43 percent of the total nitrogen in mulberry.

Palatability

One of the main features of mulberry as forage is its high palatability. Small ruminants avidly consume the fresh leaves and the young stems first, even if they have never been exposed to it before. Then, if the branches are offered unchopped, they may tear them off and eat the bark. Cattle consume the whole biomass if it is finely chopped. There is a report of ad libitum DM intake of 4.18 percent of liveweight (average of three lactating goats), which is much higher than in other tree fodders. Jayal and Kehar reported DM intakes of mulberry leaves of 3.44 percent of body weight in sheep under experimental conditions. Animals initially prefer mulberry over other forages when they are offered simultaneously, and will even dig through a pile of various forages to look for it.

Mulberry Planting

The most common planting method worldwide is by stem cuttings, but in certain places seed is preferred. As is the case with other tropical perennial forages for cut-and-carry systems, planting by seed assures deeper roots with a greater capacity to find water and nutrients that eventually results in higher biomass production and greater longevity. Seeds might be the most acceptable way of transporting, quarantine and store selected materials. The advantages of stem reproduction (cloning) are certainty of production characteristics, practicality in obtaining material and ease of planting.

Male plants might be preferred when introducing foreign germoplasm to new locations since this prevents involuntary expansion. As in most perennial forages, the time and the establishment cost (mainly for land preparation, planting and weed control) are critical for the successful introduction of mulberry.

Mulberry is cultivated for fruit as isolated trees or in orchards; for small-scale silkworm rearing along boundaries or along food crops in mixed farming systems; for large silk projects or for intensive forage production in pure stands; and also for forage in association with N-fixing legumes. Mulberry is also found mixed with other trees in natural forests or plantations. All the required nutrients for mulberry growth must come from the soil, since mulberry does not fix atmospheric nitrogen.

In pure stands, mineral and organic fertilisers (animal and vegetable manures) must be used to replenish the nutrients removed with the foliage in order to maintain a sustainable production. The association with legumes with effective N-fixing rhizobium can reduce N inputs and may be the most desirable combination for some farms, but even when recycling nutrients in animal manures, extra chemical fertilisers are required for maximum yields. Responses of mulberry to N fertilisers have been clearly demonstrated, both in inorganic and organic forms, with better responses to the latter (Table 5). The nitrogen level in soils is the major factor for mulberry growth.

Table 5. Effect of goat manure or ammonium nitrate application on total DM yields during three consecutive years

Year	Level of manure (tonnes DM/ha/year)				NH_4NO_3
	0	240	360	480	480
1	23.0	24.4	26.6	31.1	26.7
2	21.3	25.2	27.6	33.4	29.7
3	22.9	28.2	32.6	38.2	29.2

For silkworm feeding, individual leaf picking, shoot harvesting and whole branch cutting are practised, depending on the feed requirements of the silkworm larvae stages and harvesting costs. For silkworms, leaves are offered fresh some other forms of feeding are being developed. For ruminant feeding, the preferred method has been branch cutting by hand, although mechanical harvesting could be employed in the future for direct feeding of fresh material on a large scale, for processing or for drying. Forage conservation by ensiling has been successfully achieved and there have been some preliminary studies on leaf drying.

Leaf blades dry within hours under full sun but more time is required for petioles and stems. Some conditioning (e.g. passing through rollers) may help to reduce water content and minimise the deterioration of leaf quality by over exposure. Diploid varieties dry more quickly since they tend to have more stomata per unit of leaf area.

The production of leaf and total DM per hectare of mulberry depends on the variety, the location, plant density, fertiliser applications and harvesting techniques. Table 6 gives the yields of mulberry in various locations.

Table 6. Examples of mulberry yields

Location	Variety	Fraction	Yield (tonnes/ha/year)	
			Fresh	DM
Karnataka, India	M-5	Leaf	40	
		Stem	52	
Mpwapwa, Tanzania	Local	Leaf		8.5
		Stem		14.1
		Bark		2.7
San José, Costa Rica	Tigreada	Leaf and young stem		11.0
	Indonesia	Leaf and young stem		8.7
Puntarenas, Costa Rica	Tigreada	Leaf and young stem		13.4
	Indonesia	Leaf and young stem		12.5
Matanzas, Cuba	Tigreada	Total biomass	30	
	Acorazonada	Total biomass	33	
	Indonesia	Total biomass	26	
	Local	Total biomass	30	
Cuyutla, Guatemala	Local	Total biomass	37	
		Leaves		16
Zhenjiang,	Shin	Leaves		32
		Ichinose		
Jiangsu, China		Branches	28	
		Stems	8	
Kalimpong, W. Bengal, India	Local	Leaves	22	
	BC 259	Leaves	20	
	TR 10	Leaves	19	
	C 763	Leaves	19	

Total biomass yield and leaf proportion vary with species and varieties. Climate (moisture and solar radiation) and soil fertility are determining factors on productivity. Increasing planting density increases leaf yields. Fresh leaf yields of 40 tonnes/ha/year (approximately 10 tonnes of dry matter) have been reported in India and in Costa Rica. Maximum dry matter yields of edible material (leaves and young stems) and total biomass were 15.5 and 45.2 tonnes/ha/year, respectively. Total leaf DM yields of less than 10 tonnes could be expected under less intensive production.

Nutritional Value

The nutritive value of Mulberry is one of the highest found in products of vegetable origin, far superior to traditional forages like alfalfa. Mulberry biomass is remarkable due to one characteristic which is found in very few plants; high levels of CP and high levels of digestible energy. It also notable for good mineral content and above all, its low fibre content.

Table 7. Dry matter, crude protein and IVDMD of Mulberry foliage and other feeds used in Central America

Item	DM,%	CP,%	IVDMD,%
Mulberry *(M. alba)*	28.7	23.0	80.0
King grass *(P. purpureum x P. typhoides)*	20.0	8.2	52.7
Star grass *(Cynodonlnemfluensis)*	22.3	8.9	54.9
Commercial concentrate	91.5	17.7	85.0

The content of dry matter (DM) and other components in the leaves of Mulberry is higher when compared to traditional grasses used in animal feeding (Table 7). From

Costa Rica there are reports of 25-32% DM in leaves; 23-29% for young stems; and 24-45% for woody stems. In plantations of three Mulberry varieties planted at 0.40m between plants and 1m between rows, in three ecologically different sites in Costa Rica and with various fertilisation levels, it was found that DM content of leaves and edible stem were more affected by the location than by fertilisation level, without differences among varieties (Table 8).

Table 8. Site and nitrogen fertilisation effects on leaf dry matter content in three Mulberry (M. alba) varieties in Costa Rica.

Site	*Variety, % DM*			*Mean*
	Criolla	*Indonesia*	*Tigriada*	
Puriscal	31.6	32.0	30.7	31.4
Coronado	26.8	25.7	22.2	24.9
Paquera	29.0	36.0	31.9	32.3
N, kg/ha/year				
180	29.2	31.4	28.8	29.8
360	28.9	31.5	28.6	29.7
540	29.2	30.8	27.4	29.1
Mean	29.1	31.2	28.3	

Table 9. Site and nitrogen fertilisation effects on woody stem dry matter content in three Mulberry (M. alba) varieties in Costa Rica

Site	*Variety, % DM*			*Mean*
	Criolla	*Indonesia*	*Tigriada*	
Puriscal	45.2	41.9	42.9	43.4
Coronado	31.6	27.1	24.2	27.6
Paquera	48.8	47.8	48.8	48.5
N, kg/ha/year				
180	42.7	39.2	39.8	40.6
360	41.8	39.9	38.4	40.0
540	41.2	37.8	37.8	38.9
Mean	41.9	38.9	38.6	

In the woody stems, the differences were still more pronounced, with values between 27 and 48% of DM (Table 9). The small effect of fertilisation on DM was also reported in a plantation where various levels of goat manure were applied (Table 10).

Table 10. Dry matter content of Mulberry (Morus alba) by soil manure application

Fraction % de MS	Manure level				NH_4NO_3
	0	240	360	480	480
Leaves	26.4	25.9	26.0	25.5	25.6
Young stem	28.0	27.1	26.8	26.0	24.7
Woody stem	42.2	41.5	40.8	40.4	40.0

Mulberry leaves have high levels of CP and IVOMD when compared to other feed used in ruminant feeding. Data from Central America indicate CP values between 15-25% and IVOMD between 75-90%, which indicates a quality comparable or superior to commercial concentrates (Table 11). Non lignified stem also has a good nutritional quality, with CP values of 7-14% and IVOMD of 56-70%

Table 11. Site and nitrogen fertilisation effect on crude protein in three Mulberry (Morus alba) varieties in Costa Rica

Site	Variety, % PC			Mean
	Criolla	Indonesia	Tigriada	
Puriscal	20.1	22.2	21.0	21.1
Coronado	22.2	26.3	25.7	24.8
Paquera	17.0	14.8	13.7	15.1
N kg/ha/year				
180	18.9	20.8	20.0	19.9
360	20.5	20.8	19.9	20.4
540	20.0	21.6	20.6	20.7
Average	19.8	21.1	20.1	

As it was the case for DM, CP and IVDMD is similar among varieties and is not affected much by fertilisation nor by cutting frequency (Tables 12, 13, 14, 15, 16 and 17). Although, an effect was observed on CP content of leaves and young stem when ammonium nitrate was added instead of goat manure at iso-nitrogenous levels (Table 15).

Table 12. Site and nitrogen fertilisation effect on crude protein of young stem in three Mulberry (Morus alba) varieties in Costa Rica

Site	Variety, % PC			Mean
	Criolla	Indonesia	Tigriada	
Puriscal	11.2	13.0	12.1	12.1
Coronado	13.4	13.9	14.3	13.9
Paquera	9.4	10.2	8.7	9.4
N, kg/ha/year				
180	11.2	12.7	11.6	11.4
360	11.8	11.8	11.8	11.7
540	11.3	12.6	11.7	11.9
Mean	11.3	12.4	11.7	

Table 13. Site and nitrogen fertilisation effect of leaf IVDMD in three Mulberry (Morus alba) varieties in Costa Rica.

Site	Variety, % PC			Mean
	Criolla	Indonesia	Tigriada	
Puriscal	78.0	75.5	76.8	76.7
Coronado	75.4	73.8	75.6	74.9
Paquera	71.6	71.2	71.6	71.5
N, kg/ha/year				
180	74.9	72.6	75.0	74.2
360	74.8	73.6	74.1	74.2
540	75.3	74.2	74.9	74.8
Year	75.0	73.5	74.7	

Table 14. Site and nitrogen fertilisation effect on IVDMD of young stem in three Mulberry (Morus alba) varieties in Costa Rica.

Site	*Variety, % PC*			*Mean*
	Criolla	*Indonesia*	*Tigriada*	
Puriscal	69.5	70.5	63.9	68.0
Coronado	70.3	68.6	69.9	69.6
Paquera	65.9	64.7	59.4	63.3
N, kg/ha/year				
180	68.5	68.2	65.4	67.4
360	68.9	69.0	64.2	67.4
540	68.2	66.6	63.7	66.2
Average	68.6	67.9	64.4	

Table 15. Manure application effect on crude protein content of Mulberry leaf and young stem

Fraction	*Manure level*				*NH_4NO_3*
	0	*240*	*360*	*480*	*480*
Leaves	19.1	19.3	19.3	20.2	22.5
Young stem	7.4	7.1	7.2	7.4	12.3

Table 16. Manure application effect on dry matter digestibility of Mulberry leaf and young stem

Fraction	*Manure level*				*NH_4NO_3*
	0	*240*	*360*	*480*	*480*
Leaves	76.7	77.5	77.0	76.9	77.1
Young stem	56.2	56.4	56.0	55.8	58.6

Table 17. Annual cutting frequency effect on digestibility and crude protein content of Mulberry biomass

Fraction	% CP		% IVDMD	
	3	4	3	4
Leaves	19.2	20.9	77.2	76.9
Young stem	8.1	8.5	56.9	56.2

In the studies mentioned above, a marked site effect is reported, which is the result of the different soil and climatic conditions of each one. In Paquera, located in the Pacific Coast of Costa Rica, with high luminosity and high temperatures, leaf CP and IVDMD are reduced compared to higher locations with more clouds and lower temperatures, like is the case of Coronado and Puriscal, located in the mountainous areas of the country. The greater luminosity and higher temperature in Paquera can explain the lower water content in all fractions, lower CP and IVDMD levels, and high Dm production, as will be seen later. It is known that high luminosity reduces nitrate levels and increases cell wall components and growth due to larger photosynthetic activity. Coronado has lower temperatures and more cloudiness and rainfall, which could have limited growth and lignification. In addition to the climatic factors of Paquera, the lower N, copper and zinc soil contents could limit fertility and nutrient content in the plant. On the other side, the clay soils, the lower pH and the low potassium contents can explain low biomass yields in Puriscal.

Working with goats fed exclusively on Mulberry and "Amapola" leaves confined to metabolic cages, apart from high DM intake values, high levels of *in vivo* DM and CP digestibilities of Mulberry leaves (Table 18). The values reached 90% for leaf CP digestibility

Table 18. Dry matter and crude protein in vivo digestibilities of Mulberry and "Amapola" (Malvaviscus arboreus) with goats in metabolic cages.

Forage species	*Parameters*		
	Intake, % LW	*IVDMD, %*	*IVCPD, %*
Mulberry	4.18	79.3	89.5
Amapola	4.35	64.2	54.2

Although statistical comparisons have not been made, nitrogen, potassium and calcium contents of leaves and young stems are high (Tables 19 and 20), reaching values of 3.35, 2.0 and 2.5%, respectively. In other work (Table 21), no appreciable differences were found in mineral content of leaves and young stems when increasing amounts of legume leaves were added to the soil.

Table 19. Fertilisation effect on the mineral content of Mulberry (Morus alba) leaf in Costa Rica

Variety	*Minerals, % of Dry Matter*				
	N	*P*	*K*	*Ca*	*Mg*
Criolla	3.17	0.30	2.07	1.90	0.47
Indonesia	3.37	0.30	1.73	2.87	0.63
Tigriada	3.23	0.40	2.33	2.74	0.55
N, kg/ha/year					
180	3.18	0.33	1.96	2.54	0.55
360	3.26	0.33	2.05	2.49	0.55
540	3.32	0.33	2.12	2.48	0.55
Mean	3.26	0.33	2.04	2.50	0.55

Table 20. Fertilisation effect on mineral content of Mulberry (Morus alba) young stem in Costa Rica

Variety	*Minerals, % of Dry Matter*				
	N	*P*	*K*	*Ca*	*Mg*
Criolla	1.81	0.30	2.24	1.33	0.39
Indonesia	1.98	0.40	2.99	1.53	0.49
Tigriada	1.87	0.33	2.41	1.38	0.40
N, kg/ha/year					
180	1.89	0.33	2.48	1.43	0.44
360	1.86	0.33	2.60	1.40	0.43
540	1.90	0.37	2.55	1.42	0.41
Mean	1.89	0.34	2.55	1.41	0.43

In a report of the proximate analysis of Kanvas-2 variety of Mulberry (Table 22), levels of protein (29.6%), ash (7.53%) and nitrogen-free extract (50.0%) were considered high. At the same time, low levels of crude fibre were reported (10.1%). In this work, the type and quantity of leaf amino acids were determined. Twenty four amino acids in significant concentrations were detected, plus 6 more at low concentrations. The low fibre level and high contents of CP and IVDMD justify future evaluations of Mulberry as an ingredient for high quality meal and compounded feeds.

Table 21. Mulberry leaf mineral content as affected by level of Poro foliage added to the soil

Mineral	*Control without trees*	*Proportion of Poro foliage in the soil*			*Mean*
		0%	*50%*	*100%*	
N	2.90	3.16	3.10	3.09	3.06
P	0.22	0.22	0.22	0.23	0.22
K	1.51	1.37	1.35	1.71	1.49
Ca	1.64	1.82	1.74	1.77	1.74
Mg	0.42	0.44	0.41	0.38	0.41

Table 22. Leaf composition of Mulberry variety Kanvas-1, in El Salvador, Central America.

Leaf position	*Fractions, % of Dry Matter*					
	Dry matter	*Crude protein*	*Crude fat*	*Crude fibre*	*Ash*	*NFE*
Terminal	19.5	33.6	3.2	9.4	7.7	46.1
Intermediate	22.5	28.3	2.8	10.2	7.3	51.4
Basal	23.0	26.7	3.4	10.8	7.7	52.6
Average	21.7	29.6	3.1	10.1	7.5	50.0

Animal Response

Evaluations carried out with ruminants show high intakes levels of DM and high animal responses in weight gain and milk yield. Oviedo when comparing Mulberry foliage with concentrate (Table 23) as supplements to Jersey x Criollo grazing cows, found similar milk yields with both supplement at equal DM intake levels and superior to grazing only. Mulberry inclusion did not affect fat, protein and total solids contents in milk (Table 24), but improved net benefits when compared to the concentrate.

Table 23. Milk yield and dry matter intake of Jersey x Criollo cows grazing Star grass and supplemented with Mulberry or concentrate

Feed consumed, % LW	*Supplement*		
	Concentrate	*Mulberry*	*Nothing*
Milk, kg/animal/day	12.45	12.08	10.34
Concentrate	1.0		
Mulberry		1.0	
Star grass	3.7	3.4	3.7
Total	4.7	4.4	3.7

Table 24. Milk chemical composition from grazing cows supplemented with Mulberry or concentrates

Component	*Supplement*		
	Concentrate	*Mulberry*	*Nothing*
Fat	3.95	3.81	3.63
Protein	3.53	3.51	3.31
Total solids	12.53	12.39	12.31

Esquivel *et al.*, when replacing 0, 40 and 75% of the concentrate by Mulberry foliage, did not find significant differences ($p<0.05$) in milk production in Holstein cows grazing Kikuyo grass (*Pennisetum clandestinum*) without appreciable effects on milk quality (Table 25). Also in this work, considering only feeding costs, net income per animal was 11.5% higher with the maximum level of Mulberry than that obtained with concentrate only.

Table 25. Effect of the substitution of concentrate by Mulberry foliage on milk yield of Holstein cows grazing Kikuyo grass (Pennisetum clandestinum)

Parameter	Relation Concentrate:Mulberry		
	100/0	60/40	25/75
Milk, kg/animal/day	14.2	13.2	13.8
Intake, kg DM/animal/day			
Concentrate	6.4	4.2	1.9
Mulberry	0	2.8	5.5
Grass	9.3	7.8	.6.2
Total	15.7	14.8	13.6

With cattle, attractive liveweight gains have been obtained when using Mulberry foliage as supplement. In the humid tropics of Turrialba, Jersey x Criollo heifers grazing Star grass (*Cynodon nlemfuensis*) and supplemented with concentrate, Mulberry and concentrate or only Mulberry, no statistically differences were detected (p<0,05) among supplements (Table 26).

Table 26. Intakes and liveweight gains of grazing dairy heifers supplemented with concentrate and Mulberry foliage

Parameter	*Only concentrate*	*Concentrate + Mulberry*	*Mulberry*
Liveweight gain	0.620	0.742	0.600
Intake of Mulberry, %LW	-	0.5	1.0
Intake of concentrate, % LW	1.0	0.5	-

The combination with Mulberry and concentrate gave the highest gains. With young Romo-sinuano bulls in total confinement and fed a basal diet of Elephant grass (*Pennisetum purpureum*), gains of 40, 690, 940 y 950 g/animal/day were observed with whole Mulberry DM intakes of 0, 0.90, 1.71 and 2.11% of LW as supplement. In this study, the benefit/cost relations were 0.10, 1.11, 1.18 and 0.97 for each of gain levels, respectively. The study lasted 70 days and animals were between 13-16 months old, with initial liveweight between 118 and 250 kg (Table 27).

Table 27. Effect of Mulberry supplementation on intake and liveweight gains of Romo-sinuano cattle in confinement and fed a basal diet of Elephant grass.

DM intake, % LW1	*Mulberry dry matter offered, % LW*			
	0	*1.0*	*1.9*	*2.8*
Fresh Mulberry	0.00	0.90	1.71	2.11
Elephant grass	2.04	1.79	1.29	0.95
Total diet	2.04	2.69	3.00	3.06
Weight gain, g/animal/day	0,04	0,69	0,94	0,95

With crossbred dairy goats of 40kg liveweight, Rojas and Benavides found milk yield increases of 2.0 and 2.5 kg/animal/day when Mulberry supplementation was raised from 1.0 to 2.6% of LW on DM basis (Table 28). There were slight increases of fat, protein and total solids contents (Table 29). In this study, high DM intakes and additive effect of Mulberry supplementation were observed. King grass was clearly substitute by Mulberry (Table 30).

With Black Belly lambs receiving a basal diet of King grass, liveweight gains of 60, 75, 85 and 101 g/animal/

day were reported when Mulberry was given as supplement at 0, 0.5, 1.0 and 1.5% of LW on DM basis, respectively. In this study, rather than a substitution effect, there was an additive effect of Mulberry on total DM intake (Table 31).

Table 28. Effect of Mulberry supplementation on milk yield (kg/animal/day) of dairy goats.

Production level	*Level of Mulberry DM, % LW*				*Mean*
	1.0	*1.8*	*2.6*	*3.4*	
High	2.04	2.38	2.51	2.47	2.35
Low	1.64	1.82	1.91	2.12	1.87
Mean	1.84	2.10	2.21	2.29	

Table 29. Effect of Mulberry supplementation on fat, protein and total solid contents of goats fed King grass

Fraction	*Level of Mulberry DM, % LW*				*Mean*
	1.0	*1.8*	*2.6*	*3.4*	
Fat	3.1	3.2	3.2	3.3	3.2
Protein	3.3	3.4	3.4	3.5	3.4
Total solids	10.7	11.0	11.4	11.2	11.1

Table 30. Effect of Mulberry supplementation on dry matter intake of confined dairy goats

Forage	*Dry matter intake, % LW*			
Mulberry	1.0	1.8	2.6	3.4
King grass	3.2	2.9	2.6	2.1
Total	4.2	4.7	5.2	5.5

Table 31. Performance of Black Belly lambs fed various Mulberry levels

Parameters	Mulberry level, DM as %of LW				Significance
	0	0.5	1.0	1.5	
Initial weight, kg	15.7	15.8	15.8	15.1	
Final weight, kg	21.5	22.6	24.4	25.6	
Gain, g/day	60	75b	85b	101	**
DM intake (kg/day)					
King grass	0.66	0.60	0.62	0.60	
Mulberry leaf	0	0.09	0.19	0.28	
Total DM intake	0.66	0.69	0.81	0.88	**
DM intake, % LW	3.54	3.72	3.99	4.34	**
DM intake, g/kg/$PV^{0.75}$	73.5	75.1	84.5	92.2	**

In a three-year evaluation (Table 32), in an agroforestry model with goats fed exclusively with King grass and Mulberry, a lactation yield of 900 kg per 300 days was reported. This is equivalent to a mean of 3 kg/day and to 4.1 kg/day at the beginning of lactation. Forage came from a Mulberry and grass plantation associated with Poro (*Erythrina poeppigina*) measuring 1,100 m^2, fertilised with goat manure, Poro foliage and feed rejects.

Table 32. Milk yield (kg/animal/day) in dairy goats fed exclusively grass and Mulberry foliage in the agroforestry model

Month	1	2	3	4	5	6	7	8	9
Goat 1	3.22	3.46	3.47	3.41	2.65	2.69	2.23	2.44	2.53
Goat 2	3.41	3.93	3.53	3.44	2.91	2.67	2.68	1.86	1.71

During the third year, the module reached 5.0 kg/day, equivalent to 16,500 kg/ha/year (Table 33). The economic analysis indicated a benefit/cost relation of 1.27, 1.39 and 1.45 for each year, respectively (Table 34).

Ensiled Mulberry

One of the most serious problems of livestock husbandry in the tropics is the rapid decline of grass quality during the dry season. Among the most used alternatives is the conservation of forage by ensiling during the rainy season in order to use it in the dry season. However, silage is traditionally made with tropical grasses rich in fibre and low in soluble carbohydrates, which affects fermentation and results in a low quality product. Due to its low fibre and high level of carbohydrates, Mulberry foliage can be ensiled without additives, showing a lactic fermentation pattern and low CP losses while maintaining between 66 ad 71% IVDMD. These parameters are far superior to silages made with tropical forages. Vallejo, using 40kg plastic bags and three 30-day storage periods, compared the silage made from three tree foliages. Mulberry showed the highest levels of IVDMD, good CP content, acceptable loses of ammonium nitrogen and higher lactic acid levels (Table 35). Acetic and butyric acids levels were also high with Mulberry in the first period, but they rapidly declined (Table 36).

Table 33. Zootechnical parameters of confined goats in the agroforestry model in Turrialba, Costa Rica.

Parameter Average	*Year 1*	*Year 2*	*Year 3*	
Lactation, months	14.6			
Kids per parturition	2.0			
Kid weight at birth, kg	3.7			
Milk,, kg/animal/day	2.0			
Dry period, days	37.3			
Goat weight, kg	50.0			
Total milk, kg/year		1223,0	1505,9	1110,5
Milk production, kg Per module/day	4.0	3.4	4.2	5.0

Maximum monthly	42.8	71.4	122.7
Minimum monthly	145.1	189.1	207.9
Minimum per module/day	1.4	2.3	4.0
Maximum per module/day	5.2	6.1	6.7

Table 34. Financial analysis (in US$) of the goat agroforestry module at Turrialba, Costa Rica.

Description	*Years*		
	1991/92	*1992/93*	*1993*
A. Costs			
A.1 Investments			
Mulberry x Poro plantation	4.61	4.61	2.88
Grass x Poro plantation	1.66	1.66	1.04
Facilities	16.07	16.07	9.37
Animals	50.00	50.00	31.25
Subtotal	72.34	72.34	45.21
A.2 Fixed, land opportunity cost	21.17	21.17	13.23
A.3 Variables (labour)			
Cut&carry, weeding, pruning	182.65	176.19	109.77
Forage chopping and feeding	138.45	133.55	83.20
Milking	89.05	85.90	53.52
Cleaning facilities	54.60	52.67	32.81
Manure application	26.00	25.08	15.63
Mineral salts	30.66	30.66	19.16
Antelmintic	1.40	1.40	1.40
Maintenance	6.50	6.27	3.90
Subtotal	455.31	511.72	319.39
Total cost	527.65	584.06	377.83
Updated cost	610.82	643.92	396.72
B. Income			
B.1 Milk production	672.66	813.99	549.03
Updated income	778.68	897.42	576.48
C. B - A updated	167.86	253.50	179.76
B/C	1.27	1.39	1.45

Table 35. Chemical characterisation of silage made from shrubs and trees

Species	*% DM*			*% IVDMD*			*% CP*		
Period	*1*	*2*	*3*	*1*	*2*	*3*	*1*	*2*	*3*
Mulberry	22.1	28.7	34.3	60.5	61.6	76.9	17.9	17.0	16.7
Amapola	28.9	30.9	35.7	51.1	55.5	68.5	16.3	16.0	16.9
Jocote	20.6	22.0	19.8	49.3	58.0	74.2	17.0	17.5	15.9

Table 36. Fermentation indicators of ensiled foliage of three woody forages

Variable	*Period*	*Species*		
		Mulberry	*Amapola*	*Jocote*
pH	1	5.4	5.0	2.9
	2	4.6	5.0	3.1
	3	5.1	5.0	2.8
% Total NH_3-nitrogen	1	2.27	0.96	1.00
	2	1.56	0.81	1.28
	3	1.26	0.16	0.91
% Acetic acid	1	5.71	1.49	1.11
	2	1.97	1.25	1.57
	3	0.85	0.16	1.10
% Butyric acid	1	6.84	0.24	0.03
	2	0.56	0.24	0.01
	3	0.32	0.02	0.00
% Lactic acid	1	5.14	4.64	0.71
	2	9.00	4.26	1.84
	3	6.22	1.38	0.63

González *et al.*, in two 9 ton Mulberry silage pits for feeding young bulls, found better fermentation indicators than Vallejo. They noted high levels of lactic acid and low concentrations of propionic and butyric acids (Table 37). pH remained close to the values reported before by Vallejo and, when opening the silo, the organoleptic characteristics were excellent: green colour and lactic smell. Nevertheless, once in the feeder the silage lost within two hours its characteristic colour and smell.

Table 37. Fermentation indicators of Mulberry silage used as supplement for Romo-sinuano cattle

Indicator	*Silo 1*	*Silo 2*	*Average*
pH	4.45	5.25	4.85
NH_3, % of total N	7.70	4.16	5.93
Acetic acid, %	3.95	1.82	2.89
Propionic acid, %	0.27	0.08	0.18
Butyric acid, %	0.02	0.03	0.03
Lactic acid, %	12.93	17.15	15.04

The use of this Mulberry silage to supplement young bulls fed a basal diet of Elephant grass, liveweight gains of 600g/day were obtained (Table 38) with silage intake of 1.1% of LW on DM basis. However, although Mulberry silage intake was similar to that observed with other grass silages, there was a lot of refusal. The low intake was not expected due to CP and IVDMD contents. The explanation could be in the physical and chemical aspects of the silage, like high pH and high compaction which might have reduced consumption. Thos forage without additive does not reach the desirable pH levels to assure stability. Gonzalez observed this in a study with micro-silos, where after 60 days the pH never dropped below 4.5.

Table 38. Effect of Mulberry silage supplementation on intake and weight gain of Romo-sinuano cattle fad a basal diet of Elephant grass.

Dry matter intake	*Mulberry dry matter offered, % LW*			
	0	*0.8*	*1.7*	*2.5*
Ensile Mulberry, % LW1	0	0.66	1.05	1.11
Total dry matter, % LW	2.16	2.42	2.61	2.64
Weight gain, g/animal/day	117	404	490	601

The conclusions that can be reached from the use of silage of woody forages relate to the species used. This observation also comes from a study with dairy goats fed silage made from three woody forages, where a very highMulberry silage intake was found and a milk yield of 1.9 kg/day.

Agronomy

Mulberry also shows high biomass production, as shown in several experiments where the effects on yield and quality of cutting frequency, level and type of fertilisation and location have been study. Due to its high production capacity, Mulberry extracts a lot of nutrients from the soil. This is why, a combination of organic and inorganic fertilisers has been emphasised.

One of the obvious worries on the agronomy of Mulberry is related stake planting. For this purpose an experiment was conducted to compare three woody forages and different planting methods. In all cases Mulberry showed a high germination percentage, superior to other forages. Placing stakes in vertical position was better than horizontally.

In an evaluation of cutting height of isolated 2 year-old plants (Table 39), differences in DM and CP production were no found, although there was a significant effect on the proportion of leaf. In this same study (Table 35), greater yields were observed with longer cutting frequencies. However, this increase was due to more stem production, with leaf production no being affected.

Table 39. Dry matter yields (kg/plant/year) of Mulberry at two cutting heights.

Plant fraction	Cutting height		p <
	0.5 m	1.0 m	
Total	2.32	2.16	NS
Leaf	1.00	1.02	NS
Stem	1.32	1.14	NS
Leaf fraction	0.21	0.21	NS
Leaf:stem	1.04	1.29	**

Table 40. Dry matter yields (kg/plant) of Mulberry under different cutting frequencies

Fraction	Cutting frequency, days			p <
	60	120	180	
Total	1.64	2.17	2.86	**
Leaf	1.11	1.04	0.84	**
Stem	0.52	1.08	2.01	**
Leaf fraction	0.26	0.23	0.15	**
Leaf:stem	2.11	1.06	0.45	**

Under humid tropical condition in Costa Rica, for three years a Mulberry plantation with 25,000 plants/ha, has produced 35 ton DM/ year using goat manure as fertiliser. It was also observed 20% higher production with goat manure than with ammonium nitrate, at equal levels of nitrogen application (Table 41). Significant increments in production were observed with greater manure applications. Overall yields increased 10% a year, reaching 38 ton of DM/ha by the third year. Greater total biomass yields were obtained (Table 42) with a 4 month cutting frequency compared with 3 month, although leaf yield was not different.

Table 41. Yearly total Mulberry biomass (ton of DM/ha) when applying goat manure as fertiliser and cutting every 4 months.

Year	Level of manure				NH_4-NO_3
	0	240	360	480	480
1	23,0	24,4	26,6	31,1	26,7
2	21,3	25,2	27,6	33,4	29,7
3	22,9	28,2	32,6	38,2	29,2

Table 42. Dry matter yields of Mulberry (ton/ha/year) per year and cutting frequency

Fraction	Year			Cuttings / year	
	1	2	3	3	4
Leaves[1]	10.8	10.9	10.4	11.0	10.4
Young stem	1.6	0.9	0.8	1.0	1.3
Woody stem	11.8	13.2	14.4	16.0	10.3
Total	24.2	25.0	25.6	28.0	21.9
Edible	12.4	11.8	11.2	12.0	11.6

In another study, always under hum2id tropical conditions, Oviedo found Mulberry biomass yields of 8.0, 9.4 and 10.6 ton of DM/ha when using Poro foliage as mulch equivalent to 0, 160 and 300 kg of N/ha/year, respectively. Working with three Mulberry varieties in three sites of Costa Rica, Espinoza *et al.* reported important differences among varieties in total DM yields, with one variety with almost half of the yields of the other two. Nitrogen application also affected yields according to site (Table 43). In Paquera, despite the long dry period and less fertile soils, average yields for all three varieties were almost double those of Coronado, where it rains all year. Edible biomass yield differences were smaller (Table 44). In Paquera lower leaf:stem proportion was observed (Tables 45 and 46) at similar cutting frequencies. This indicates that more frequent cuts can be given in Paquera conditions.

Table 43. Site and nitrogen fertiliser effect on dry matter production of three Mulberry varieties in Costa Rica.

Parameter	Variety, ton DM/ha/year			Average
	Criolla	Tigriada	Indonesia	
Site:				
Puriscal	11.1	15.6	19.0	15.2
Coronado	8.9	19.5	18.0	15.5
Paquera	22.4	31.9	39.2	31.2
Fertilisation				
180 kg N/ha	11.2	18.0	19.2	16.1
360 "	13.7	22.8	28.3	21.6
540 "	17.4	26.3	28.7	24.1
Average	14.1	22.3	25.4	

Table 44. Site and nitrogen fertiliser effect on edible dry matter percentage of three Mulberry varieties in Costa Rica

Parameter	Variety			Average
	Criolla	Tigriada	Indonesia	
Site				
Puriscal	53.8	50.8	38.6	47.7
Coronado	64.7	56.4	48.5	56.5
Paquera	40.0	42.0	32.0	38.0
N, kg/ha/year				
180	52.8	51.5	41.3	48.5
360	52.8	49.1	38.6	46.8
540	52.9	48.7	39.6	47.1
Average	52.8	49.7	39.7	

Table 45. Site and nitrogen fertiliser effect on leaf:stem relation in three Mulberry varieties in Costa Rica

Parameter	Variety			Average
	Criolla	Tigriada	Indonesia	
Site				
Puriscal	1.00	0.56	0.89	0.82
Coronado	1.40	0.71	0.87	0.99

Paquera	0.61	0.44	0.67	0.57
N, kg/ha/year				
180	1.03	0.60	0.87	0.83
360	0.99	0.54	0.81	0.78
540	1.00	0.57	0.75	0.77
Average	1.00	0.57	0.81	

Table 46. Site and nitrogen fertiliser effect on growth rates (ton DM/ha/year) of three Mulberry varieties in Costa Rica

Parameter	*Variety*			*Average*
	Criolla	*Tigriada*	*Indonesia*	
Site				
Puriscal[1]	92.5	130.0	158.3	126.9
Coronado	74.2	162.5	150.0	128.9
Paquera	186.7	265.8	326.7	259.7
Fertilisation				
180 kg N/ha	93.3	150.0	160.0	134.4
360 "	114.2	190.0	235.8	180.0
540 "	145.0	219.2	239.2	201.3
Average	117.8	186.1	211.7	

Ruminants

Although the feeding value of mulberry for dairy cattle has been recognised for some time in Italy and it has been traditionally used in Himalayan countries, research on mulberry for ruminants has been rather limited. Jayal and Kehar, given on the high digestibility values of M. indica leaves, suggested that they could be used as supplements for lower-quality forages. Mulberry was used to replace grain-based concentrates in lactating cows with excellent results (Table 47). Yields did not significantly decrease when 75 percent of the concentrate was replaced with mulberry.

Table 47. Substitution of concentrates by mulberry in lactating Holstein cows grazing Kikuyu grass (Pennisetum clandestinum)

Parameter	Concentrate: Mulberry		
	100:0	60:40	25:75
Milk yield (kg/d)	14.2	13.2	13.8
Intake (kg MS/d):			
Concentrate	6.4	4.2	1.9
Mulberry	0	2.8	5.5
Kikuyo grass	9.3	7.8	6.2
Total	15.7	14.8	13.6

Milk production increased with the levels of mulberry offered to goats on a King grass diet as shown in Figure 1. At CATIE, Turrialba, Costa Rica, a module of two dairy goats (Saanen x Toggenburg), fed exclusively with forage from 775 m² of mulberry (17 000 plants/ha), in association with Erythrina berteroana (5 128 trees/ha) just as green manure, and from 425 m² of King grass, produced an average of four litres per day, equivalent to over 12 000 litres per ha/year.

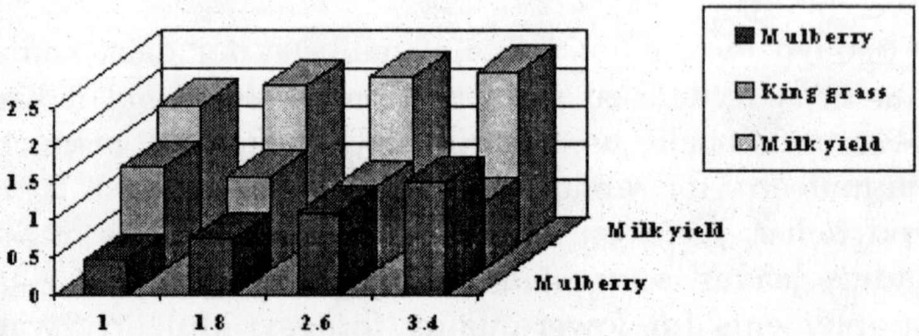

Figure 1. Milk yield and forage intake in goats fed increasing levels of mulberry on a King grass diet

Also in Costa Rica, liveweight gains of bulls belonging to the Romosinuano breed (a criollo type) fed Elephant grass, increased to over 900 g/day when mulberry was offered as a supplement at 1.7 percent of their body weight on a DM basis. Table 48 presents the results of an experiment in Guatemala with growing Zebu x Brown Swiss steers being fed increasing levels of mulberry as a supplement to a sorghum silage diet.

Table 48. Effect of mulberry supplementation level on intake and weight changes of Zebu x Brown Swiss steers fed on sorghum silage

Parameter	Mulberry level (% BW)			
	0	0.5	1.0	1.5
DM intake (percent BW/day				
Total	2.26	2.39	2.64	2.88
Sorghum silage	2.26	1.91	1.68	1.51
Daily gains (g/d)	- 128	- 29	164	195

BW = Body weight.

Although the growing rates with the highest mulberry level are not impressive (195 g/day), most likely because of the poor quality of the silage, this trial shows the high nutritive value of the supplement. Total intake and weight changes improved with the amount of mulberry offered reflecting its higher nutritive value compared to the basal diet. Daily gains of female calves (0-4 months) were not affected when mulberry leaves were offered ad libitum and the commercial concentrate reduced to 25 percent of the amount traditionally used. In lambs, gains reached 100 g/day when King grass was supplemented with 1.5 percent dry matter of mulberry.

Monogastrics

The silkworm has a relatively simple digestive system. In some ways it is comparable to that of monogastric animals, so that, in theory, mulberry leaves could also be used at least as one of the ingredients in monogastric diets. In a trial with growing pigs in which a commercial concentrate was replaced by up to 20 percent of mulberry leaf, the best level of substitution was 15 percent. It increased daily gains from 680 g/day ith only concentrates, to 740 g/day and also gave the best economic results.

In rabbits, the reduction of concentrate offered daily from 110 to 17.5 g with ad libitum fresh mulberry only reduced gains from 24 to 18 g/day, but decreased to more than half the cost of the meat produced. The combination of mulberry and Trichantera gigantea leaves, as the protein source, and blocks made of molasses, cassava root meal and rice bran, as the energy source, gave better reproduction and growth performance than a diet of commercial concentrates and grass. Singh, Goel and Negi supplemented Angora rabbits, receiving pelleted diets, with mulberry leaves ad libitum and obtained intakes of mulberry equivalent to 29-38 percent of the total intake. This level significantly reduces feed cost.

Deshmukh, Pathak and Takalikar fed mulberry leaves as the sole ration for adult rabbits. These authors found daily intakes of 68.5 g for dry matter, 11.2 g for crude protein and 175 kcal for digestible energy (equivalent to 2.55 Mcal of digestible energy per kg). The digestibility values were 74 percent for crude protein, 59 percent for crude fiber and 64 percent for dry matter. The authors concluded that mulberry leaves provided enough nutrients for maintenance. Narayana and Setty found better egg yolk colour and increased egg size and production with the inclusion (up to 6 percent) of shade-dried M. indica leaf meal in the mash of laying hens.

Other small herbivores, such as guinea pigs, iguanas and snails, could also be fed mulberry leaves. In fact, wild green iguanas (Iguana iguana) came to feed on recently established mulberry fields in Costa Rica.

Livestock Production Systems

The traditional way of using mulberry as animal feed in silkproducing areas is by providing ruminants with the residue left by the silkworm. A model of sericulture and milk production has been proposed by Mehla, Patel and Tripathi, in which dairy cows receive mulberry residue and concentrates. The generation of edible protein and employment are much greater than with food grains. This refuse material is added to fish ponds for herbivorous carps in the Chinese dyke-pond system, which is one of the most intensive agricultural low-input systems in the world, and generates food and outputs for a large number of people. In these silk areas, as well as where mulberry grows wild, cut-and-carry systems are practised and it is the most obvious way of utilising mulberry for livestock, either from pure stands or from associations with legumes.

Mulberry foliage can constitute the supplement to low-quality forage (grass) based diets or as the main component of the ration. A natural association of mulberry and livestock occurs in regions such as the Near East and Central Asia where mulberry trees are kept for fruit production. Fallen leaves in the autumn are consumed by domestic animals. Since fruit ripens in late spring or early summer, it may be possible to harvest leaves for forage one or more times before the winter. The only suggestion of utilising mulberry for direct grazing came from FAO, proposing a complementary association with clover (Trifolium subterraneum) for sheep and cattle grazing in Tuscany (Italy). Mulberry benefits from the N fixation by the clover and contributes with high-quality forage during the summer. The association produces more forage over a longer period than the individual pure stands.

Trees and Shrubs for Ruminant Feeding

The use of trees and shrubs for ruminant feeding has been practised among producers in Central America for decades. This empirical knowledge about forage properties of various species is of great value for science. In several studies to characterise production systems, producers report a large number of species for browsing and for cut-and-carry systems with animals in confinement. Studies on the subject have been oriented towards valuation of trees and shrubsas a source of forage and their integration into ruminant production systems.

The focus has been on agroforestry and farming systems, and the aim has been to develop technological alternatives allowing more sustainable production and rational use of soil and forest resources. In order to consider a tree or a shrub as forage, it must have advantages above traditional forages from the point of view of its nutritional value, its yield or its agronomic versatility. To qualify, these requirements are:

i) nutritional content and intake that will allow animal performance improvements;

ii) resistance to repeated pruning/harvesting; and

iii) high biomass yields per unit area.

Apart from these features, it is advisable to select native species to take advantage of their adaptation to the environment, and species that can easily be established with simple and inexpensive techniques. Direct observation of animals eating has allowed to identification species that are particularly palatable and have high digestibility (in vitro) of the organic matter (IVOMD) and high crude protein (CP). These studies have permitted species without current use to be valued, together with others normally used for different purposes.

The information provided by producers has also allowed simple and easy to implement agronomic management practices to become known. Examples are the woody forages identified in the western Plateau of Guatemala where, in most cases, propagation is carried out by vegetative means (stem cuttings) with which a faster biomass can be obtained compared with sexual seed. The forage from most of these woody forages shows CP values two and three times higher than tropical grasses and, in some cases, even higher than commercial concentrates used to supplement ruminants.

At the same time, the IVOMD is very high and comparable with or superior to that of concentrates. Two species of the Euforbiaceae family are outstanding in nutritional quality, Wide Chicasquil (Cnidoscolus acotinifolius) and Fine Chicasquil (C. chayamansa. Their leaves, with more than 30 percent CP and 75 percent IVOMD, are edible. Also outstanding in nutritive value are two species of the Moraceae family, mulberry (Morus spp.) and Amate (Ficus spp.) from Petén, Guatemala; two from the Malvaceae family, the Amapola (Malvaviscus arboreus) and the Clavelón (Hibiscus rosa-sinensis); Black Sauco (Sambucus mexicana) and Yellow Sauco (S. canadensis), belonging to the Caprifoliaceae family; and three species of the Compositeae family, Chilca (Senecio spp.), White Tora (Verbesina turbacensis) and Purple Tora (V. myriocephala). All of these have CP higher than 20 percent and IVOMD higher than 70 percent.

Animal Performance

In intake trials, Poró (Erythrina poeppigiana), was the most studied species in the 1980s, with observed values higher than 4 percent with lactating goats. In other work, acceptability has been sought in forage species growing in semi-arid lands, under forests and in secondary forests, which were identified by means of direct observation of grazing animals. Both Poró and Black Wood (Gliricidia sepium), are legumes characterised by high CP content, but with lower IVOMD. In these cases, research has shown that energy supplementation significantly improves animal performance and that starch sources give a better response than simple sugars.

With species with higher nutritive value, the highest milk yields have been obtained and a significant response has been observed when the foliage level is increased on grass-based diets. That is the case with Amapola and Mulberry, with milk yields in goats of 2.2 and 4.0 kg/d, normally only possible with commercial concentrates. With these species, intakes higher than 5 percent of body weight are reported. With mulberry foliage, increasing weight gains have been observed when raising its proportion in the diet. Of the known technologies, vegetative propagation is the most used, since it shortens the establishment period, is easy to carry out and widely known by producers.

Germination percentage exceeds 95 percent, when stem cuttings of mulberry and Amapola are used in humid tropical conditions. With the Yellow Sauco, nursery planting of stakes and further transplanting seem to be the most suitable propagation method. In some species it is possible to place stakes horizontally, obtaining several plants and sparing material. However, there are important variations among species, which are important to know before making a decision about which technique to use. The association of leguminous trees and grasses is a viable alternative with two modalities.

In the first, the forage of both grass and tree is utilised. Results of an experiment under humid tropical conditions, using King grass (Pennisetum purpureum x P. typhoides) with Poró with no nutrient replenishment and all biomass removed showed that grass yields are not reduced by the presence of the tree, since tree pruning reduces competition for light. It was also found that digestible nutrient yields per area were triple compared with the grass alone. Nevertheless, in the short term for the grass and in the medium term for Poró, productivity declines because of lack of nutrient replenishment. The other way is to utilise associated Poró foliage as green manure for the grass.

In the humid tropics, grass yields improved when increasing amounts of Poró foliage were applied. At the same time, the sole presence of the tree, even without pruning, stimulated grass growth compared with grass in a monoculture without trees. Traditionally in livestock husbandry, there is an unidirectional relationship between animals and plants, the first benefiting by obtaining the feed, but without contributing to its generation. In confined systems, it is possible to establish a two-way relationship, since most manure can be used as fertiliser. In this way a more balanced system can be established and the plants will benefit from the animal excreta. This particularly applies to those woody species and the best forage features.

With their high biomass yields and without being able to fix nitrogen, they require high levels of fertiliser applications. In order to find a rational ecological solution, goat manure has been applied to Amapola and mulberry plantations. The yields have been 18 and 30 tonnes of dry matter per ha, respectively. With mulberry, yields increase over years. Of great importance, in sites with bimodal rainfall pattern, are tree-pruning techniques that allow abundant biomass production during the dry season. The effects of pruning at the end of the dry period have been studied.

In the Dominican Republic, pruning of Black Wood (Piñon cubano or Gliricidia) in the months of October, November and December, in addition to stopping flowering, results in increasingly higher biomass yield during the dry period. Similar results were obtained by other authors working with Amapola and Jocote (Spondias purpurea). To date, most of the technologies have been implemented in small farms and in goat production systems oriented towards family consumption.

In these situations, in addition to the technological aspects, it is essential to know the economic viability of the alternatives developed, both at the station and on the farm. For the economic aspects, partial budget analysis of experiments; profitability analysis (flow and net income) of implemented technologies in pilot modules; and analysis of family benefits and flow and net income at the farm level have been used. The analysis, indicates that the application of forage tree technologies on the farm is profitable and contributes to improving family economy. With dairy goats under a basal diet of grass, the use of Poró foliage and other agricultural by-products (e.g. reject bananas) as a supplement is more profitable than the use of concentrates, despite higher yields with the latter.

Total cost of DM, from planting to feeding in some forage species nutritionally comparable with commercial concentrates, is lower. This in part explains the high profitability of a dairy goat model at CATIE with goats fed exclusively mulberry and grass. At the level of family subsistence units, high profitability has been found when family labour is not considered, even with animal reproductive problems present.

3

Mulberry Cultivation and Utilisation

Mulberry foliage is the only food for the silkworm (*Bombyx mori*) and is grown under varied climatic conditions ranging from temperate to tropical. Mulberry leaf is a major economic component in sericulture since the quality and quantity of leaf produced per unit area have a direct bearing on cocoon harvest.

Table 1. Area under mulberry in different states

State	*Area (ha)*
Andhra Pradesh	38 084
Assam	2 813
Jammu & Kashmir	4 717
Karnataka	166 000
Kerala	1 164
Madhya Pradesh	2 043
Manipur	25 975
Tamil Nadu	9 491
Uttar Pradesh	5 665
West Bengal	21 358
Other	4 934

In India, most states have taken up sericulture as an important agro-industry with excellent results. The total area of mulberry in the country is around 282 244 ha. The details of the area under mulberry cultivation in different states in India is shown in Table 1.

Cultivation of Mulberry Species

There are about 68 species of the genus *Morus*. The majority of these species occur in Asia, especially in China and Japan. Continental America is also rich in its *Morus* species. The genus is poorly represented in Africa, Europe and the Near East, and it is not present in Australia.

In India, there are many species of *Morus*, of which *Morus alba, M. indica, M. serrata* and *M. laevigata* grow wild in the Himalayas. Several varieties have been introduced belonging to *M. multicaulis, M. nigra, M. sinensis* and *M. philippinensis*. Most of the Indian varieties of mulberry belong to *M. indica*. In China there are 15 species, of which four species, *Morus alba, M. multicaulis, M. atropurpurea* and *M. mizuho* are cultivated for sericulture. In the former Soviet Union *M. multicaulis, M. alba, M. tartarica* and *M. nigra* are present.

Though mulberry cultivation is practised in various climates, the major area is in the tropical zone covering Karnataka, Andhra Pradesh and Tamil Nadu states, with about 90 percent. In the sub-tropical zone, West Bengal, Himachal Pradesh and the northeastern states have major areas under mulberry cultivation. The details of the mulberry varieties under cultivation in different states of India is given in Table 2.

Mulberry is a fast growing deciduous woody perennial plant. It has a deep root system. The leaves are simple, alternate, stipulate, petiolate, entire or lobed. The number of lobes varies from one to five. Plants are generally dioecious. Inflorescence is catkin with pendent or drooping peduncle bearing unisexual flowers. Inflorescence is always auxiliary. Male catkins are usually longer than the female catkins. Male flowers are loosely arranged and after shedding the pollen, the inflorescence dries and falls off. These are four persistent parianth lobes and four stamens implexed in bud. Female inflorescence is usually short and the flowers are very compactly arranged.

There are four persistent parianth lobes. The ovary is one-celled and the stigma is bifid. The chief pollinating agent in mulberry is wind. Mulberry fruit is a sorosis, mainly violet black in colour. Most of the species of the genus *Morus* and cultivated varieties are diploid, with 28 chromosomes. However, triploids (2n=(3x)=42) are also extensively cultivated for their adaptability, vigorous growth and quality of leaves.

Climatic Requirements

Mulberry thrives under various climatic conditions ranging from temperate to tropical located north of the equator between 28° N and 55°N latitude. The ideal range of temperature is from 24 to 28°C. Mulberry grows well in places with an annual rainfall ranging from 600 to 2 500 mm. In areas with low rainfall, growth is limited through moisture stress, resulting in low yields. On average, mulberry requires 340m^3/ha of water every ten days in case of loamy soils and 15 days in clayey soils. Atmospheric humidity in the range of 65-80 percent is ideal for mulberry growth. Sunshine is one of the important factors controlling growth and leaf quality. In the tropics, mulberry grows with a sunshine range of nine to 13 hours a day. Mulberry can be cultivated from sea level up to an elevation of 1 000 m.

Table 2. Mulberry varieties in India

Variety	*Region*	*Developed at*	*Origin*
Kanva-2	South India Irrigated	CSRTI, Mysore	Selection from natural variability
S-36	South India Irrigated	CSRTI, Mysore	Developed through EMS treatment of Berhampore Local
S-54	South India Irrigated	CSRTI, Mysore	Developed through EMS treatment of Berhampore Local
Victoria-1	South India Irrigated	CSRTI, Mysore	Hybrid from S30 x Berc 776
DD	South India Irrigated	KSSRDI, Thalaghattapura	Clonal selection
S-13	South India Rainfed	CSRTI, Mysore	Selection from polycross (mixed pollen) progeny
S-34	South India Rainfed	CSRTI, Mysore	Selection from polycross (mixed pollen) progeny
MR-2	South India Rainfed	CSRTI, Mysore	Selection from open pollinated hybrids.
S-1	Eastern and NE India Irrigated	CSRTI, Berhampore	Introduction from (Mandalaya) Myanmar
S-7999	Eastern and NE India Irrigated	CSRTI, Berhampore	Selection from open pollinated hybrids
S-1635	Eastern and NE India Irrigated	CSRTI, Berhampore	Triploid selection
S-146	N. India and Hills of J and K Irrigated	CSRTI, Berhampore	Selection from open pollinated hybrids
Tr-10	Hills of Eastern India	CSRTI, Berhampore	Triploid of Ber. S1
BC-259	Hills of Eastern India	CSRTI, Berhampore	Back crossing of hybrid of Matigare local Kosen with Kosen twice
Goshoerami	Temperate	CSRTI, Pampore	Introduction from Japan.
Chak Majra	Subtemperate	RSRS, Jammu	Selection from natural variability
China White	Temperate	CSRTI, Pampore	Clonal selection

Soil Condition

Mulberry flourishes well in soils that are flat, deep, fertile, well drained, loamy to clayey, and porous with good moisture holding capacity. The ideal range of soil pH is 6.2 to 6.8, the optimum being 6.5 to 6.8. Soil amendments may be used to correct the soil to obtain the required pH. The quantity of gypsum or lime to be applied in different cases to bring the pH to 6.8 are given below:

Gympsun

pH range	*Gypsum/ha*
7.4 to 7.8	2.0 tonnes
7.9 to 8.4	5.0 tonnes
8.5 to 9.0	9.0 tonnes
9.1 and above	14.0 tonnes

Lime

pH range	*Lime/ha*		
	Plain	*Hilly areas*	*Soil type*
5.5 to 6.5	1.25 tonnes	2.5 tonnes	Sandy
	2.50 tonnes	5.0 tonnes	Sandy loamy
	5.0 tonnes	7.5 tonnes	Loamy
	7.5 tonnes	8.75 tonnes	Clay loamy

Method of application: The powdered gypsum/lime is mixed well with garden of soil and irrigated to stagnation for 48-72 hours. Later the water is leached out by drainage and dried.

Mulberry Cultivation under Rainfed Regions

Suitable Mulberry Varieties

Kanva-2, S-13 and S-34 varieties are recommended for

rainfed (rainfall: 500-800 mm) regions of South India (Karnataka, Andhra Pradesh and Tamil Nadu).

Kanva-2: Belongs to *Morus indica.* Diploid. Widely cultivated in Southern India. Selection from natural population of Mysore local variety. Inflorescence and sorosis: female, profuse flowering, many soroses. Production characteristics: medium leaf maturity, yields about 30 to 35 tonnes/ha/year under irrigated conditions. Leaf moisture content 70 percent, protein content 21 percent and sugar content 11.5 percent. High rooting ability (80 percent) and wide adaptability. Resistant to leaf spot. Moderately resistant to leaf rust and powdery mildew.

S-13: Belongs to *M. indica.* Selection from open pollinated hybrids of Kanva-2. Recommended for rainfed areas of South India during 1990. Inflorescence: male, profuse flowering. Production characteristics: yields 8-12 tonnes/ha/year under rainfed conditions, depending on rainfall. Moisture content 70.6 percent protein content 24.3 percent and sugar content 13.8 percent. Resistant to leaf spot and powdery mildew, moderately resistant to leaf rust and tukra infestation.

S-34: Belongs to *M. indica* Diploid. Selection from progeny of S30 x Berc 776. Recommended during 1990 for rainfed areas with black cotton soils of South India. Inflorescence and sorosis: male, profuse flowering, occasionally few soroses. Production characteristics: under rainfed conditions, yields about 15 tonnes/ha/year. Moisture content 70 percent, crude protein content 23.7 percent. Soluble sugar content 13.2 percent. Resistant to powdery mildew and leaf rust. Moderately resistant to leaf spot and susceptible to tukra infestation.

Cultivation

Land Preparation: Land for mulberry cultivation is

ploughed deep with a heavy mould board plough up to a depth of 30-35 cm. Thereafter the land is repeatedly ploughed two or three times with a country plough to bring the soil to a fine tilth. The land should be properly levelled. A basal dose of welldecomposed farmyard manure (FYM) or compost is applied at the rate of 10 tonnes/ha and thoroughly incorporated into the soil.

Spacing: The spacing commonly followed for a rainfed garden is 90 x 90 cm Pits of 35 x 35 cm are prepared. About 1 kg FYM/pit should be added.

Stakes and planting: Branches of 8-10 months old and about 50 mm in diameter should be used for the preparation of stakes of 22-25 cm length with five to six healthy buds. Three stakes are planted per pit in a triangular form with a spacing of 15 cm, leaving only one bud exposed above soil surface. If planting is done with saplings, then one sapling is sufficient perpit. Planting should be done during June/July after the onset of the monsoon.

Intercultivation: During the first year, intercultivation should be done manually. Once mulberry plants are established, bullock ploughing is carried out.

Fertilisation: 50N:25P:25K (kg/ha/year) in two doses. First dose: suphala (15:15:15) 167 kg, after 2 months of planting. Second dose: urea 55 kg or cam (100 kg) or ammonium sulfonate (125 kg), at end of September or early October before cessation of monsoon rains.

Pruning and leaf harvest: The first crop should be harvested six months after plantation when the mulberry is well established. Two more crops are harvested during the first year by the leaf picking method. Mulberry should be pruned after one year at the onset of the next monsoon. Pruning is done by sharp sickle or pruning saw at a height of 25-30 cm from the ground.

Green manuring and mulching: Green manure crops can be grown as an intercrop with mulberry during the monsoon only. Green manure crops (cowpea, horse gram, dhaincha) should be incorporated into soil by ploughing before the flowering starts and well before the rains cease. Subsequently, plots may be mulched with any dry material or plants that will not cause needs.

Maintenance

Recommended inputs (per ha per year) for gardens maintained under rainfed conditions at a spacing of 90 cm x 90 cm:

- FYM or compost, 10 tonness in a single dose at the onset of monsoon
- Azotobacter biofertiliser, 4 kg/crop, twice a year (during rainy season)
- VAM inoculum, 1 000 kg, once in mulberry lifespan (inoculation through maize rootlets)
- Suphala, 167 kg, first crop
- Single super phosphate, 156 kg, first crop
- Muriate of potash, 42 kg, first crop
- Urea (55 kg) or cam (100 kg), third crop
- Green manuring, 15 kg

Crops such as horse gram, cowpea, sun hemp and *dhaincha* should be incorporated into the soil by ploughing before flowering and cessation of the monsoon.

Leaf harvest: Individual leaf harvesting should be carried out. The expected yield (tonnes/ha/year) for different varieties is: Kanva-2, 10-12; S-13, 14-15; S-34, 14-15.

Cultivation under Irrigated Regions

Suitable Mulberry Varieties

Kanva-2, S-36, S-54, DD, MR-2 (especially in Tamil Nadu) and Victoria-1 varieties are recommended for irrigated conditions.

Kanva-2: Belongs to *M. indica.* Diploid. Widely cultivated in southern India after it was recommended for cultivation in 1969 by CSRTI (Mysore). Selection from natural population of Mysore local variety. Inflorescence and sorosis: Female, profuse flowering, many soroses. Production characteristics: Medium leaf maturity, yields 30-35 tonnes/ha/year under irrigated conditions. Leaf moisture content 70 percent, protein content 21 percent and sugar content 11.5 percent. Resistant to leaf spot. Moderately resistant to leaf rust and powdery mildew. High rooting ability (80 percent) and wide adaptability.

S-36: Belongs to *M. indica.* Developed at CSRTI and recommended during 1984. Evolved from Berhampore local by chemical mutagenesis. Cultivated in southern India. Moderate rooting ability. Production characteristics: yields 38-45 tonnes/ha/year under assured irrigated conditions of South India. Moisture content 76 percent, CP 22 percent and carbohydrate content 28 percent. Tolerant to leaf spot and powdery mildew. Moderately susceptible to leaf rust and to tukra infestation. Most suited to young age silkworm rearing. Sapling plantation recommended because of moderate rooting ability.

S-54: Belongs to *M. indica.* Developed at CSRTI and recommended during 1984. Selected from Berhampore local by chemical mutagenesis (EMS). Recommended for assured irrigated conditions of South India. Production characteristics: yields about 45 tonnes/ha/year under assured irrigated conditions. Highly responsive to agronomical input. Moisture content of leaf 70.5 percent, CP 23.9 percent and sugar content 13.8 percent. Moderately resistant to powdery mildew and leaf rust and resistant to leaf spot. Loses moisture very quickly. Good rooting ability.

DD: Selected from natural population of Dehra Dun variety and recommended by Karnataka State Sericultural Research and Development Institute, Thalaghattapura. Recommended for southern India. Morphology: erect, thin branches, coarse leaves, greenish grey bark. Lower branches spreading, leaves unlobed, big size, ovate shape. Yields 35-40 tonnes/ha/year under assured irrigation conditions.

MR-2: Belongs to *M. sinensis.* Diploid. Selection from open pollinated hybrid population. Developed at CSRTI and recommended for propagation in Tamil Nadu. Mainly cultivated in Tamil Nadu under both irrigated conditions in the plains and rainfed conditions in hilly regions. Production characteristics: yields 30-35 tonnes/ha/year under irrigated conditions of Tamil Nadu. Moisture content 68 percent, protein content 23.2 percent, sugar content 13.2 percent (Fig. 7). Resistant to powdery mildew disease. Suitable for hilly areas.

Victoria-1 (V-1): Belongs to *M. indica.* Recently developed from a cross of S-30 and Berc 776 at CSRTI. Recommended during 1996 for assured irrigated conditions. Flower: male, profuse flowering, occasionally few soroses. Production characteristics: yields about 70 tonnes/ha/year under assured irrigated conditions. Very high sprouting. Moisture content 78.9 percent and 72.5 percent in young and matured leaves respectively, protein content 24.6 percent and total sugar content 16.98 percent. Moderately resistant to leaf rust and tukra infestation and resistant to leaf spot. Quick sprouting ability and very high rooting ability (> 94 percent) high photosynthetic rate and higher water use efficiency are additional advantages. Moreover, leaves are suitable for both young and grown bivoltine silkworm rearing.

Cultivation

Selection of site: Mulberry flourishes well in soils that are flat, deep, fertile, well drained, loamy to clayey, and porous with a good moisture holding capacity. The ideal range of soil pH is 6.2 to 6.8. Mulberry can be grown in saline, alkaline and acidic soils after suitably amending the soils.

Preparation of land: Land for mulberry cultivation is ploughed deep with heavy mould board plough up to a depth of 30-35 cm. Thereafter the land is repeatedly ploughed tw or three times with a country plough to bring the soil to a fine tilth. The land should be properly levelled. A basal dose of welldecomposed FYM or compost is applied at the rate of 20 tonnes/ha and thoroughly incorporated into the soil.

Planting: Plantations can be raised by using both cuttings and saplings. The varieties ideally suited for irrigated conditions are Kanva-2, S-36 and V-1. Branches of 6-9 months old and about 15 ml in diameter should be used for the preparation of cuttings of 15-18 cm in length with three to four healthy buds for raising in the nursery or for planting directly in the field.

Spacing: Plant spacing of 90 x 90 cm is ideal for mulberry. Two cuttings per pit to be used for direct planting. When using saplings, only one sapling/pit is necessary. A paired row plantation with spacing of (90 + 150) cm x 60 cm, is recommended.

*Inter-cultivation:*Two months after planting, weeding is carried out. A second weeding is done after another two to three months. Thereafter, intercultivation should be carried out after every shoot or leaf harvest.

Irrigation: The plantation should be taken up during the onset of the monsoon to take advantage of the rain. If the rain is not sufficient, the land should be irrigated at regular intervals of 8-14 days depending on the type of soil. About one and a half to two acre inches of water is required/irrigation.

Fertilisation: The total dose of fertiliser to be applied in the first year is 100 N: 50 P: 50 K/kg/ha/year. This is applied in two doses. The first dose is applied when the plantation is about two months old at the rate of 50 N: 50 P: 50 K/kg/ha. The second dose is applied after leaf harvesting at the rate of 50 kg N/ha.

Pruning: After six months of plantating, mulberry attains a height of 1.5 to 1.75 m and is ready for harvest. The first harvesting is by bottom pruning. The second leaf harvesting is 12 weeks from the first leaf harvest and the third harvest 12 weeks from the second harvest by shoot harvest. From the second year onwards, harvesting is done at an interval of 70 days by the shoot harvest method.

Maintenance

Spacing: 90 x 90 cm or (90 + 150) cm x 60 cm

Recommended inputs:

- FYM (20 tonnes) in two equal doses with one tonne of vermicompost
- Azotobacter (20 kg) in five equal doses
- N-triacontanol (250 ml) in two equal doses (sprays): first between 10-15 days after pruning, second, ten days after thefirst spray

- VAM inoculum (1 000 kg) One dose in plantation of lifespan (inoculation through maize as the host the plant). Not required for plantations established with saplings inoculated with mycorrhiza
- Ammonium sulphate (750 kg), urea (325 kg) or calcium ammonium nitrate (600 kg) in five equal doses
- Single super phosphate (375 kg) in two equal split doses (crops 1 and 3)
- Muriate of potash (200 kg) in two equal split doses crops 1 and 3 days after first spray

Leaf harvest: Leaves are harvested either by individual leaf harvest or shoot harvesting. The latter is more economical and used for the shoot method of silkworm rearing. The expected yield (tonnes/ha/year) of varieties is Kanva-2, 32-35; S-36, 38-45; and Victoria-1, 60-70.

Cultivation in Hilly Areas

Suitable Varieties

S-1, S-7999, S-1635, S-146, Tr-10 and BC-259 varieties are recommended for the hilly regions of north and northeastern India.

Cultivation

Land preparation: If the land has a gentle slope, it can be levelled by minor land shaping and providing suitable type of bunds across the slope. If the slope is greater, contour bunding terrace planting or contour line planting can be adopted. In more sloping areas, platforms for individual plants on contour lines are more suitable since this involves less soil cutting.

Spacing: Spacing for tree planting depends on soil topography, the extent of land available for cultivation and training method. For gentle slopes, 3′ x 3′, 5′ x 5′ may be adopted. In sloping more land 10′ x 10′ can be adopted. Pits are to be prepared for plantation. In deep textured loose soils, 45 x 45 cm and in hard shallow soils 60 x 60 x 60 cm pits are to be prepared. For each pit, 5 kg of FYM or compost must be applied.

Planting: Saplings of five months age with five to six roots are suitable for planting during the regular onset of the monsoon. One sapling per pit should be planted. The saplings should be supported with a stick to ensure straight growth.

Maintenance: After one month, all the buds except the top five to six should be removed carefully without damaging the bark. Weeds around the plant should be removed and regular pot watering given. After three months of plantings second weeding should be done and 25 g of suphala/plant should be applied in a trench and should be covered with soil. A second dose of fertiliser (25 g urea/plant) should be applied before cessation of the monsoon. Plants must be protected from grazing.

Recommendations for Hilly Areas (Tropical High Land)

Spacing: 90 x 90 cm.

Recommended inputs

- FYM/compost (20 tonnes) in one single dose in January or February
- Azotobacter (7.5 kg) in three equal doses
- VAM inoculum (1 000 kg) in a single dose in lifespan plantation.
- Calcium ammonium nitrate (200 kg) in three equal doses
- Single super phosphate (156 kg) in two equal doses
- Muriate of potash (84 kg) in two equal doses

Since the recommendations are general, quantities of fertilisers and amendments may be applied on the basis of soil test reports.

Cultivation in Temperate and Sub-temperate Regions

The varieties Goshoerami and China white are suitable for temperate. Chak majra and S-146 for sub-temperate regions. For Kashmir, 20 tonnes/ha/year of FYM should be applied under irrigated conditions (bush or dwarf tree) after the annual pruning in July-August, and 300:150:150 of NPK in two equal doses in April-May and in June-July. The expected leaf yield of these varieties is: Goshoerami and China White 15-20 tonnes/ha/year and for Chak Majra 20-22 tonnes/ha/year.

Mixed Farming

Mulberry can be successfully grown as an intercrop (medium mixed tree) between rows of tea/coffee as shade plants. Besides providing shade, a substantial quantity of leaves can be obtained for silkworm rearing and for feeding cattle and goats. Furthermore, the pruned shoots are a good source of firewood. Mulberry can also be grown as an intercrop for cultivation in coconut plantations. A survey has shown that mulberry is intercropped with coconut in the areas of Channapatna, Ramanagaram, Kanakapura and Bangalore in the state of Karnataka.

Pesticide use

The following products should be applied per ha/year, if needed, at the recommended doses:

- DDVP (Nuvan), 1.25 litres (in 480 litres of water), spray, to control tukra and leaf roller. Washing soap (non detergent), 5 kg mixed with DDVP solution

— Ladybird (CSRT), 625 adult beetles as biological control of mealy bug

— Bionematicide (shelf life of 180 days), 80 kg split in three doses every four months during intercultivation operations to control root knot disease

— Neen oil cake, two tonnes in four split doses every three months during intercultivation operations to control root knot disease

— Raksha, 1 kg for 100 plants with 50 kg FYM for root rot disease

— Bavistine, at 0.2 percent, 1 kg dissolved in 500 litres of water and sprayed to prevent leaf spot and powdery mildew diseases

— Kavash, 1 kg dissolved in 500 litres of water and sprayed to prevent leaf rust

— Dithane M-45 1 kg dissolved in 500 litres of water and sprayed to prevent leaf blight due to fungi and bacteria

— Zinc sulphate 4.4 kg in 440 litres of water to increase leaf quality and quantity

— Glyphosphate at 41 percent, 8.1 litres in two split doses (2 and 4 crop) to control weeds.

Animal Feeding Practices

Mulberry is known in India as "Kalpa Vraksha" as all the parts of the plant have many uses. It is essential to sericulture as the foliage constitutes the sole feed of the mulberry silkworm. Mulberry is a fast-growing tree which, for the convenience of sericulture practices, is maintained as a bush. It produces very large amounts of renewable biomass in the form of branches, shoots, leaves and fruit. If mulberry is used for silkworm rearing it is possible to obtain 30-35 tonnes/ha of leaf every year. By growing mulberry, a farmer obtains fodder, fuel and fertiliser. With regard to fodder for animals, farmers in India feed their cows and goats with leftover branches and leaves from silkworm rearing.

Many farmers feed their animals with surplus foliage but always mix it with straw. Farmers also use the mulberry branches for fuel after pruning. Leftover twigs are allowed to dry in the garden itself. Residues of rearing are also converted to valuable FYM for mulberry gardena by putting them in a pit for four to five months prior to use. As mulberry is mainly propagated by cuttings in the tropics and sub-tropics, a certain quantity of pruned branches can be used for the preparation of cuttings and the remainder as fuel. One hectare of mulberry garden yields about 12.1 tonnes of mulberry sticks. The energy generated/ha is 27 830 Kcal. Mulberry could be exploited as an "energy crop" in cultivable, wasteland, low-lying areas, canal bunds, by roadsides and at fringe areas of the forest, etc. under various afforestation, watershed development and soil conservation programmes.

The uses of the various species of the genus *Morus* are enumerated below:

M. laevigata

The trees of this species produce sweet fruits that are used in juice and jam making in central India. In northeast India the wood is utilised as firewood; in house building and furniture making; for making stocks, spokes, poles, shafts of carriages and casts. The wood is suitable for plywood making and panelling, carving and making of toys and tea chests. It is also used for making tennis rackets. The straight log of the tree is used as a support in house-building work.

M. serrata

The wood is used for furniture making and carving, toy making, sports goods, agricultural implements and cheap types of rifles and guns.

M. alba

This species is cultivated in the hilly and plain areas of India for silkworm rearing. It is also used as a tree in roads and in social forestry. The fruit are made into juice, liquor and stews. The wood is used in the sports goods industry. It is also used for house building; agricultural implements; furniture; for making spokes, poles, shafts and bent parts of carriages and carts. The stem bark is used for making paper.

M. indica

The cultivated forms that are utilised in silkworm rearing belong to *M. indica*. There are a few profuse fruiting varieties occurring in Maharastra and Meghalaya that can be utilised as the female parent in breeding programmes. The fruit is used for jam, jelly and juice making in Maharashtra. The pruned branches are used as fuel.

Medicinal Uses

The various parts of the mulberry plant find use in Ayurvedic preparations. The leaves have diaphoretic and emollient effects and are used for making a decoction that can be used as a gargle that throat inflammation. The fruits are used to treat sore throat, depression, high fever and are both a coolant and laxative. The root extract has hypoglycaemic properties. The root bark is used as an anthelmintic, purgative and vermifuge. Mulberry root juice is administered to patients with high blood pressure. The Chinese use the leaf tips from young leaves to boil with tea to control blood pressure. The milky latex is used as a plaster for sores and for the preparation of dermal creams.

Fruiting Habit

In South India, fruits are observed in two seasons a year: during October-November and during March-May. However, whenever mulberry is pruned or defoliate flowering takes place together with sprouting of auxiliary buds followed by fruit formation. This feature of mulberry is utilised in mulberry breeding programmes. The immature fruits are green in colour but change to purplish to violet black. In certain species such as *M. alba* the fruits are white to pinkish and very sweet. In *M. laevigata* the fruits are very long, sometimes measuring up to 18 cm.

Chemical Composition of the Leaf

This differs according to variety and maturity. However, on the basis of the analysis carried out at CSRTI (Mysore), the chemical composition of the leaf is as follows:

Component	*Range*
Moisture	65 - 78 percent
Protein	19 - 25 percent
Minerals	10 - 15 percent
Reducing sugars	1.2 - 1.9
Sugars	10 - 15 percent

Pests and Diseases of Mulberry

The following are the *Key* pests of mulberry

– *Maconellicoccus hirsutus* (mealy bug) - causing tukra in mulberry
– *Diaphania pulverulentalis* - Leaf roller
– *Spilarctia obliqua* - Bihar hairy caterpillar (sporadic pest)

Minor pests of mulberry are, thrips, jassids, scale insects, shorthorned grasshopper.

Utilisation of Mulberry in India

Mulberry can be grown successfully in all conditions, even in tropical, subtropical and temperate climates. It can survive with rainfall ranging from 400-4 500 mm per annum. Although the optimum temperature for growth is between 18 and 30°C, mulberry can survive even when the temperature goes beyond 48°C or below 0°C. Mulberry can therefore be considered a universal plant, which can grow anywhere under varied climatic conditions. Although this does not guarantee successful silkworm rearing, the plant's wide adaptability can be exploited in plantations, fully or partly, for animal feed.

The important role of green leaves in supplementing animal feed is unquestionable. In the developing countries, cereal straws and grasses are fed to animals, but they cannot support full performance because of their poor nutritive value. In most places where concentrates are also provided, the diet is not balanced. Mulberry leaf supplementation can improve the efficiency of the whole diet. In addition, because of increasing pressure from the human population and to higher incomes earned from cereal and cash crops rather than from forage crops, more agricultural land cannot be set aside for fodder production.

Another consequent advantage of mulberry is that it can be grown either as a tree or a shrub and harvested several times a year. In countries such as India, where mulberry is primarily grown for sericulture, excess leaves and leftovers are fed to cattle, sheep and goats. In hilly areas, where mulberry trees are abundant, their leaves are fed to animals. Of the different species available, *Morus alba* and other species that are suitable for silkworm rearing are cultivated in the plains and on hilly ground.

Morus serrata, Morus laevigata and *Morus australis* are grown in the hills. Because of their deep root system, the leaves remain green for most of the year, except in winter when the leaves fall. As a result of the above advantages, mulberry can be considered as a perennial source of feed for most of the year. Mulberry fodder is considered to be of good quality. Leaf yield varies with nutrient supply and irrigation. In general, under irrigation with the recommended dose of fertiliser, mulberry yields nearly 35-45 tonnes of fresh leaf/ha/year.

On a dry matter basis, leaves contain an average of 20-23 percent CP, 8-10 percent, total sugar, and 12-18 percent minerals. The leaves contain nearly 70 percent moisture. The cell wall constituents are neutral detergent fibre 45.6 percent, cell contents 54.4 percent, acid detergent fibre 35.0 percent, hemicellulose 10-40 percent, lignin 10 percent, cellulose 21.8 percent and silica 2.7 percent. The contents of minerals are as follows: magnesium, 0.52-1.25 percent; chlorine, 0.02-0.29 percent; sulphur, 0.18-0.76 percent; potassium 0.93-3.19 percent and sodium 0.13-0.23 percent. Tannin accounts for 0.85 percent in leaf DM.

The chemical composition of the leaves varies from season to season. In hilly areas where the plants are not pruned repeatedly the protein content gradually decreases from February onwards, and minimum values are found in December when leaf fall starts. Like nitrogen, phosphorus also shows the same trend, its maximum content being in February. Mulberry leaf stalks and leftovers, after silkworm feeding, are generally fed to cattle. Their composition on a DM basis is 11.5 percent CP, 34.0 percent CF, 76.5 percent total carbohydrate, 9.3 percent total ash, 1.6 percent Ca and 0.2 percent P.

The calculated values were 70.8 percent for digestible CP, 48.4 percent for total digestible nutrients and 35.6 percent for starch equivalent on a dry basis. The balance of N, Ca and P were positive and animals also gained weight. A trial feeding of mulberry leaves to sheep was conducted by Jayal and Kehar. The digestible nutrients of leaves on a DM basis were 10.7 percent for CP, 8.3 percent for CF and 39.6 percent for total digestible nutrients. Narayana and Setty indicated that incorporation of 6 percent shade-dried mulberry leaves with poultry feed increased egg production.

To ascertain the effect of feeding leftover mulberry leaves with stalks, a study was conducted by the authors during 1999-2000 at Kolar, the traditional sericultural area of Karnataka (India) with hybrid milk cows (local x Jersey) at the farmer level. Twenty farmers, who regularly feed mulberry leftover leaves and stalks after silkworm feeding, were selected. In general, only one feeding of mulberry leaves at noon was carried out. There was a control group of ten cows. On average, 15-20 kg of either full mulberry leaves or leftover leaves with stalks were fed to the cows. In the case of the control group, the same feeding schedule was followed except at noon, when mulberry leaves were replaced by cereal straw. The amount of feed supplied at different times is shown in Table 3.

Farmers provided mulberry leaves only once a day. It was found that animals did not have any reluctance in accepting mulberry as feed. Records were kept on milk yield, fat percentage and time required for next pregnancy for both the groups. The data collected were subjected to statistical analysis to study the significance of feeding mulberry leaves. The total quantity of feed was constant for both groups (Table 3). Animal performance data are presented in Table 4.

Table 3. Schedule of feeding and quantity of feed given to milk cows.

Feeding time	*Control feed*	*kg*	*Mulberry feed*	*kg*
06:00h	Fodder grass	15	Fodder grass	15
11:00h	Finger millet grass	20	Finger millet grass	20
13:00h	Rice straw	15	Mulberry	15
22:00h	Concentrate	10	Concentrate	10

Table 4. Comparison of normal feeds vis-à-vis partial mulberry feeds with respect to economic characters in hybrid cows (local x Jersey)

Group	*Milk yield (litres)*			*Fat %*	*Days to conception*
	Morning	*Evening*	*Total*		
Control (n = 10)	8.4	6.7	15.1	2.94	100
Mulberry (n = 20)	11.0	9.2	20.2	3.50	77
"t" value	9.203	9.366	9.747	21.43	19.56
Significance	P<0.01	P<0.01	P<0.01	P<0.01	P<0.01

The study clearly shows that mulberry improved feed quality significantly. Nearly 34 percent higher milk yield was recorded when mulberry was provided once a day. Milk yield was found to increase both morning and evening, although the improvement was found to be more pronounced evening milking. Fat content was higher in cows fed mulberry. Interestingly, cows fed mulberry conceived 24 days. However, 25 percent of farmers believe that continuous feeding with mulberry can cause termination of pregnancy in some cases for unknown reasons.

4

Mulberry and its Potentials for Animal Feeding

The domestication of mulberry started several thousands years ago as a requirement for silkworm rearing. Considering that the silk trade has existed for a long time throughout the Old World and that mulberry is also cultivated for its fruit and for landscaping, its germplasm has been taken to many countries, and it now has a very wide distribution range in Asia and Europe; in Africa and in the Americas. The origins of most cultivated mulberry varieties are believed to be in the China/Japan area and in the Himalayan foothills.

Silk production was important in Europe during the nineteenth and early twentieth centuries, as it was in Japan and Korea up to the middle of the twentieth century. Silk production nowadays is dominated by China and India. Although in some countries of Asia mulberry leaves have been traditionally fed to farm animals, interest in the intensive production of mulberry and its utilisation for animal production started in several countries in the late 1980s and early 1990s. The attractive biomass yields, palatability and exceptionally high nutritive value for ruminant and monogastric animals have been the reasons behind the great interest in mulberry for animal feeding in recent years.

Mulberry Species

There are about 68 species of the genus *Morus*, and the majority of them occur in Asia. In China, there are over a thousand varieties under cultivation. They originated from four main species, the White mulberry (*Morus alba*), the Lu mulberry (*M. multicaulis*), the Mountain mulberry (*M. bombycis*) and the Guangdong mulberry (*M. atropurpurea*).

In India the main species are *M. indica, M. alba, M. serrata* and *M. laevigata,* which grow naturally in the north of the country. Most of the cultivated varieties belong to either *M. indica* or *M. alba*. In Mysore, India, the Central Sericulture Research and Training Institute keeps 244 mulberry cultivars: 78 indigenous, 44 exotic, 21 unknown and 101 elite hybrids.

In the republics of the former Soviet Union, the most common species are *M. multicaulis, M. tartarica* and *M. nigra*. In Indonesia there are seven species: *M. alba* (varieties *tartanica* and *macrophyla*), *M. nigra, M. multicaulis, M. australis, M. cathyana* and *M. mierovra*. In Viet Nam there are over 100 varieties, mainly *M. alba, M. nigra* and *M. laevigata*. According to Janaki Ammal only *M. rubra* is native to the Americas.

The most popular species in the world are believed to be *M. alba* and *M. indica*. These have been the subject of intensive selection from open pollination, controlled hybridisation and selection and mutation breeding in several countries, resulting in over a thousand varieties, including many polyploids. In Brazil there are about 90 varieties, all *M. alba*.

Uses of Mulberry

Sericulture

The most important use of mulberry globally is in the production of silkworm, which feeds exclusively on its leaves. The country with the largest area of mulberry is China with approximately 626 000 ha, then India with nearly 280000 ha. Several other countries, such as Thailand and Brazil (35 000 ha), still have some mulberry production but on a much smaller scale. Silk projects have been and are being started in various developing countries, particularly in Africa and Latin America. Regardless of how successful or sustainable they have been, these projects have been responsible for the introduction and dissemination of mulberry varieties under different soil and climatic conditions.

Fruit

All throughout Asia, but particularly in Central Asia and in the Near East, mulberry is highly appreciated for its delicious fruit, which is consumed fresh, or in the form of juice or conserves. A few mulberry orchards also exist in Latin America.

Wood

Especially in the Indian subcontinent, mulberry wood is used for handicrafts, cabinet work and for sporting woods (e.g. grass-hockey sticks and tennis rackets). Its thin branches are woven into baskets.

Landscaping

In Asia, southern Europe and in the southern United States, mulberry trees are utilised for landscaping. Their resistance to pruning and their low water requirements make them very suitable plants for urban conditions, house gardens, street shade and city embellishment.

Medicine

A variety of medicinal properties has been attributed to the different parts of the mulberry plant. Leaves are also dried and used in infusions in Asia.

Forage

Silk producers have traditionally fed mulberry surpluses from silkworm feeding to farm animals and to herbivorous carp in polyculture fish ponds. In this sense, farm animals are well integrated into silk production. There have even been economic studies of the combined benefits from silk and dairy production in India. A review of mulberry for animal feeding has been published by FAO.

In countries south of the Himalayas, where mulberry trees occur in nature, their foliage has been part of the traditional mixed diet of domestic ruminants. In India there have been numerous research reports on the use of mulberry residues and leaves for various domestic animals, from cows to poultry. In general terms, Indian scientists have considered mulberry foliage as medium-quality forage.

Some small farmers in East Africa, mainly in the United Republic of Tanzania and in Kenya, harvest foliage from mulberry trees and include it as part of the diet offered to ruminants in confinement. Although its nutritive value is recognised in these countries, there has been little planting of mulberry for this purpose. Since the late 1980s and mainly in the late 1990s, the cultivation of mulberry as forage has expanded considerably in Latin America, in Central America and in the Caribbean in particular, but it is still mainly associated with research institutions. This has been mainly the result of the initial studies carried out by the Tropical Agriculture Research and Higher Education Center (CATIE) in Costa Rica and in other Central American countries and the expansion by a number of research and development projects. There have been, however, in other countries in Latin America, independent initiatives to feed mulberry leaves to various species of farm animals.

Mulberry for Animal Production

It is not by chance that mulberry germplasm is causing so much interest as an alternative high-quality feed for farm animals. Over several hundreds, and perhaps thousands, of years, mulberry species and varieties have been selected and improved to feed the silkworm, which is nutritionally very demanding. The aim has been to produce greater quantities of leaves of higher quality under a wide range of conditions.

Although, in general terms, the principles of mulberry cultivation for sericulture purposes should apply to its cultivation as forage for feeding farm animals, there are certain important differences. Frequency of harvesting and, in particular, planting density, can be intensified if the goal is to feed ruminants, since the quality is already high enough and the main purpose is to increase overall biomass yields. The issues of maintenance of soil fertility and plant persistence become important if huge quantities of nutrients are extracted from the soil in the biomass under cut-and-carry systems. Individual leaf picking, as is commonly practised for silkworm feeding, can only be justified for small-scale or family units of mini-livestock, such as snails, guinea pigs or rabbits. Mechanical harvesting might be more appropriate for ruminant feeding in larger operations.

Feeding ruminants

The urgent need for a high-quality feed for ruminants in the tropics, in particular for small ruminants, and the excellent characteristics of mulberry, are the justifications for the great enthusiasm over its intensive cultivation and use as a feed supplement for cattle, and as the main feed for goats.

The nutritional quality of locally produced mulberry leaves is equivalent to that of grain-based concentrates. Thus, they are an ideal supplement in most forage diets. The greatest impact of mulberry on livestock productivity is to be expected in the dairy cattle sector in the tropical regions.

Feeding Mini-livestock

The nutritive value of mulberry leaves becomes greater in inverse proportion to animal size, since metabolic rate and hence nutrient requirements decrease with size. Mulberry leaves should be the preferred feed for guinea pigs, rabbits and perhaps snails. The performance of dairy goats fed mulberry-based diets in Costa Rica has been impressive. Many more excellent results are to be expected when mulberry is offered to other herbivores, particularly small species.

Mulberry for Browsing

Several studies have examined the possibility of mulberry for direct browsing by cattle in Italy, in France and in Japan. Although the results have been promising, a great deal of work will need to be done before mulberry is incorporated into grazing systems on a large scale.

5

Nutritional Quality of Mulberry

The developing countries such as China have little land for ruminant production per se, as crop production for human consumption is given the first priority. It is inconceivable that large tracts of land could be set aside for fodder tree planting. The animal production system should be integrated with the whole agricultural system.

Silkworm cultivation on the leaves of the mulberry tree (*Morus alba*) is an important farming activity in southeast China, with over 10^6 and 10^5 ha planted in China and Zhejiang Province, respectively. The biomass yield of fresh leaves is of the order of 15 to 22 tonnes/ha/year. Silkworm production is not always profitable, as it is dependent upon the price of silk and world market trends in this commodity.

Alternative ways of using the foliage of mulberry tree would be welcomed by farmers for the occasions when income from silkworm is low. The leaves of mulberry are relished by sheep and goats and have high nutritive value with protein content at about 20 % of dry matter. They could be appropriate protein supplements for sheep fed a basal diet of ammoniated rice straw as partial or total replacement for oilseed meals, which could then be used in the diets for mono-gastric animals. However, little information is available on the nutritional values of mulberry leaves as animal feed.

Mulberry has adapted excellently to the tropics of Central America and gives very high yields. For forage production, mulberry has shown excellent characteristics of palatability and, consequently, high intakes by cattle. The literature reports that the whole plant contains between 14-22 percent crude protein on DM basis and *in vitro* digestibility between 70 to 80 percent.

The nutritional value of mulberry is in its leaves. However, being a shrub, biomass supply is considered an important aspect when used in animal feeding. Boschini, Dormond and Castro reported leaf:stem ratios above 1, before 100 days of growth. Leaf:stem proportion varies with cultivation and management conditions, plant density and cutting frequency. The whole mulberry plant, once chopped, is offered to cattle. This represents a mixture of leaf and stem which the animal selects during consumption. For this reason it is necessary to know precisely chemical composition of leaves and stems. The results of the influence of plant density, cutting height and frequency on the chemical composition throughout the year in a high tropical environment are presented here.

Research was conducted at the Dairy Cattle Experimental Station "Alfredo Volio Mata" at the University of Costa Rica. The station is located at 1 542 m above sea level, and has a mean rainfall of 2 050 mm, distributed during the months of May to November. From December to mid-May, there is no rainfall. Mean temperature is 19.5°C and relative humidity 84 percent. The soil is volcanic, classified as Typic Distrandepts. It is characterised by its medium depth, good natural drainage and medium fertility. The pH is 5.9. Ecologically, the zone typifies as low mountain humid forest. Agricultural practices, seeding conditions and management were described by Boschini, Dormond and Castro.

The plantation was left to grow for one year, during which weeding was practised and fertilisation with ammonium nitrate was applied at the rate of 150 kg of N/ha/year in two equal doses (July and October). The following year in May, the three plots received a standardisation cut, with scissors, half of the plot at 30 cm and half at 60 cm from the ground. From that date, cuts were given every 56 days (6 cuts), 84 days (4 cuts) or 112 days (3 cuts), during the experimental period of 336 days. After each cut, the plots were weeded, leaving the needs between rows. When the shoots reached 3-5 cm (approximately two weeks after harvest) they were fertilised with ammonium nitrate at a dose equivalent to 150 kg of N/ha/year.

Mulberry samples were collected at each spacing, cutting height and frequency. The branches were weighed fresh in the field, and then the leaves and stems were separated and weighed. Each sample was dried in the oven at 60°C for 48 hours until weight was constant. The samples were then milled and subjected to DM (at 105°C), crude protein (Kjeldall) and total ash determinations. NDF, ADF and lignin were analysed by the method described by Goering and Van Soest. Hemicellulose and cellulose values were obtained by difference. DM and chemical fractions were analysed with PROC GLM of the statistical package SAS. Variation sources with statistical significance were subjected to the Duncan test to differentiate the significance among means.

The chemical composition of leaves and stem as affected by planting density, cutting height and frequency are presented in Table 1. DM percentage in the fresh forage was very constant (P >0.05) in leaves for the different planting densities. Cutting height and frequency did not influence DM percentage (P >0.05).

Table 1. Leaf and stem chemical composition (% DM) of mulberry at three spacings, two cutting heights and three frequencies

Factor	*DM*	*CP*	*NDF*	*Hem.*	*ADF*	*Cel.*	*Lignin*	*Ash*
				Leaf				
Spacing								
60 cm	24.5	23.7	33.5	9.2	24.4	18.5	5.8	16.7
90 cm	24.5	23.5	32.4	8.3	24.1	19.1	5.1	24.5
120 cm	24.8	23.2	32.6	8.6	24.1	18.7	5.4	24.8
Height								
30 cm	24.1	23.8	33.5	9.4	24.1	18.9	5.3	17.0
60 cm	25.1	23.1	32.1	7.9	24.2	18.7	5.6	16.5
Frequency								
56 days	24.1	25.6	34.3	9.1	25.2	19.5	5.7	16.1
84 days	23.7	22.2	32.3	8.8	23.5	18.6	4.9	16.9
112 days	26.8	20.8	30.7	7.6	23.1	17.4	5.7	18.1
				Stem				
Spacing								
60 cm	23.2	8.7	63.2	14.5	48.7	40.0	8.7	8.2
90 cm	23.4	8.8	63.2	14.5	48.8	40.5	8.3	8.3
120 cm	24.9	8.8	64.4	16.3	48.1	39.6	8.5	7.7
Height								
30 cm	23.4	8.8	63.6	14.8	48.8	40.2	8.6	8.0
60 cm	24.2	8.8	63.6	15.4	48.2	39.8	8.4	8.1
Frequency								
56 days	20.9	10.7	59.0	15.7	36.6	43.3	6.7	10.0
84 days	23.6	7.8	65.9	12.7	43.2	53.2	10.0	7.1
112 days	30.0	6.3	69.8	17.0	42.7	52.7	10.1	5.5

However, the difference in DM content between 30 and 60 cm in cutting height was less than 1 percent. There were no differences in DM percentage between 56 and 84 days, and only 2.3-3 percent between them and 112 days. Crude protein content showed very little differences (P <0.05) between spacings and cutting heights. These variations were accentuated (P <0.01) between cutting frequencies, decreasing two to three percent for every 28 days from eight weeks growth.

Planting densities did not influence (P >0.05) cell walls, ADF, hemicellulose, cellulose, lignin and total ash. Cell wall and total ash showed very small differences between cutting heights. Hemicellulose, ADF, cellulose and lignin showed not significant differences (P >0.05) between cutting heights.

Cutting height influenced cell wall and total ash (P <0.01), ADF and cellulose (P <0.05). Changes in hemicellulose and lignin were not significant (P>0.05) with height. The interaction planting density and cutting frequency was significant (P <0.01) for leaf DM content and its content of protein and minerals. The structural fractions were not affected (P <0.01) by the group effect (Table 2).

In Table 3, the chemical composition of leaves and stems planted at different spacing and harvested at different heights are presented. DM, CP, cell walls, hemicellulose and total ash all had an interaction effect (P <0.05) due to these variables. ADF and its components were not affected. The associated effects of cutting height and frequency (Table 4) on chemical composition were significant for CP, cell wall (P <0.01) and total ash (P <0.01). Contents of DM, NDF components and hemicellulose did not show important variations.

Table 2. Leaf and stem chemical composition (% dry matter) of mulberry cultivated at three planting distances and harvested at three cutting frequencies.

Spacing (cm)	Frequency (days)	DM	CP	NDF	Hem.	ADF	Cel.	Lignin	Ash
				Leaf					
60	56	24.4	25.9	35.0	10.1	24.8	19.3	5.5	16.3
60	84	23.2	22.5	33.5	9.6	23.9	18.4	5.4	16.6
60	112	26.3	20.9	30.6	6.5	24.1	16.8	7.2	17.6
90	56	24.8	25.8	33.4	8.6	24.9	19.3	5.5	16.1
90	84	23.2	22.0	31.3	7.3	24.1	18.6	4.4	17.0
90	112	25.6	21.0	31.6	8.9	22.6	17.7	4.9	18.0
120	56	23.1	25.1	34.5	8.6	25.9	19.9	6.0	15.9
120	84	24.6	22.3	31.9	9.5	22.5	17.8	4.8	16.9
120	112	28.5	20.6	29.8	7.3	22.5	17.6	4.9	18.6
			Stem						
60	56	21.4	10.6	58.5	15.7	42.8	36.4	6.4	9.9
60	84	21.2	7.8	64.9	11.4	53.4	42.9	10.5	7.6
60	122	29.6	6.3	70.3	16.1	54.3	43.4	10.9	5.
90	56	21.2	10.7	58.6	14.9	43.7	37.1	6.7	10.2
90	84	23.0	7.8	66.1	12.3	53.8	44.3	9.4	7.2
90	112	28.2	6.3	68.7	16.6	52.1	42.2	9.9	5.7
120	56	20.1	10.8	59.8	16.5	43.3	36.2	7.1	9.7
120	84	26.6	7.6	66.7	14.4	52.4	42.4	10.0	6.5
120	112	32.1	6.3	70.2	18.3	51.9	42.4	9.5	5.3

Table 3. *Leaf and stem chemical composition (percent of DM) of mulberry cultivated at three planting distances and harvested at two cutting heights*

Spacing (cm)		*Height (cm)*	*DM*	*CP*	*NDF*	*Hem.*	*ADF*	*Cel.*	*Lignin*	*Ash*
					Leaf					
60	30	24.6	33.9	33.9	9.6	24.2	18.8	5.4	17.3	
60	60	24.4	23.4	33.2	8.6	24.5	18.3	6.3	16.1	
90	30	23.5	23.7	32.0	8.2	23.8	18.7	5.1	17.2	
90	60	25.6	23.3	32.8	8.3	24.4	19.4	5.0	16.4	
120	30	24.3	23.5	34.8	10.4	24.4	19.1	5.4	16.5	
120	60	25.3	22.9	30.5	6.8	23.7	18.3	5.4	17.2	
					Stem					
60	30	23.0	8.8	62.7	13.7	49.0	40.3	8.7	8.3	
60	60	23.5	8.7	63.7	15.2	48.4	39.8	8.6	8.0	
90	30	23.3	9.0	63.2	15.2	48.0	39.6	8.4	8.2	
90	60	23.5	8.7	63.3	13.8	49.5	41.3	8.1	8.3	
120	30	24.0	8.7	64.9	15.5	49.4	40.7	8.6	7.5	
120	60	25.8	8.9	63.8	17.1	46.8	38.4	8.4	7.9	

Table 4. Leaf and stem chemical composition (percent of DM) of mulberry harvested at two cutting heights and three cutting frequencies

Height (cm)	Freq. (days)	DM	CP	NDF	Hem.	ADF	Cel.	Lignin	Ash
				Stem					
30	56	23.6	25.8	35.0	10.0	25.0	19.7	5.4	16.5
30	84	23.0	22.8	34.0	10.2	23.8	18.5	5.4	17.0
30	112	26.8	21.0	29.9	7.2	22.8	17.7	5.2	18.1
60	56	24.6	25.4	33.6	8.3	25.4	19.3	6.0	15.7
60	84	24.4	21.7	30.5	7.4	23.1	18.7	4.4	16.7
60	112	26.9	20.6	31.4	8.0	23.4	17.2	6.2	18.0
				Stem					
30	56	21.5	10.7	58.9	15.2	43.7	37.0	6.7	9.9
30	84	22.0	7.9	66.0	12.3	53.7	43.4	10.2	7.2
30	112	29.2	6.2	69.7	17.2	52.6	42.4	10.2	5.4
60	56	20.3	10.7	59.1	16.2	42.9	36.2	6.7	10.0
60	84	25.2	7.6	65.8	13.1	52.7	43.0	9.7	7.0
60	112	30.7	6.4	69.8	16.9	52.9	42.9	10.0	5

Table 5 presents the values of leaf chemical components detailed by cutting number within each frequency. The effect was highly significant (P<0.01) for all chemical fractions studied. The highest level interaction, density by cutting height, influenced DM, CP and total ash (P <0.01). None of the structural carbohydrates showed important variations.

DM in stems did not show important variations (P >0.05) between planting densities. Between cutting height there was a difference (P <0.05) of less than 1 percent and between cutting frequencies there were significant variations (P <0.01). CP content was very constant (P >0.05) between densities and cutting heights. Cutting frequency influenced (P <0.01) CP very clearly. Cell wall,

ADF and its components cellulose and lignin, did not show important variations (P <0.05) between planting densities, but hemicellulose and total ash were affected slightly (P<0.05). Cutting height did not influence stem structural carbohydrates or total ash. Cutting frequency had a marked effect (P <0.01) on cell wall and its components, and on total ash.

The interaction between density and cutting frequency (Table 2) was significant (P <0.01) for stem DM but CP, ash and structural carbohydrates were not. Table 3 shows the chemical composition of the stem of plants at different densities and cutting heights. Hemicellulose, ADF and cellulose were affected by the associated action of these factors significantly (P <0.01). Other components were not affected.

When analysing the interaction of cutting height and frequency on stem chemical composition (Table 4), only DM was affected (P <0.01). Table 5 presents stem chemical composition by harvest number at each cutting frequency. The analysis showed that harvest numbers within each frequency had a significant effect (P<0.01) on all chemical components studied. The interaction between density and cutting frequency only affected stem DM content (P <0.05).

Table 5. Leaf and stem chemical composition (percent of DM) of mulberry harvested at three cutting frequencies by cut number

Freq. (days)	*Cut #*	*DM*	*CP*	*NDF*	*Hem.*	*ADF*	*Cel.*	*Lignin*	*Ash*
				Leaf					
56	1	22.0	27.9	37.4	15.3	22.0	17.8	4.2	15.4
56	2	22.0	27.6	46.0	10.2	35.8	28.6	7.2	16.7
56	3	26.6	23.9	33.9	9.8	24.1	17.9	6.3	16.0
56	4	25.6	24.5	27.8	5.4	22.4	16.5	5.9	16.6

56	5	24.0	24.1	30.6	7.4	23.2	17.5	5.6	16.8
56	6	24.4	25.7	30.2	6.5	23.6	18.7	4.9	15.0
84	1	15.8	24.7	34.5	10.4	24.1	18.7	5.4	15.8
84	2	25.0	21.1	35.7	12.7	23.0	18.1	5.1	17.9
84	3	26.2	22.1	26.9	4.1	22.9	18.5	4.4	17.3
84	4	27.8	21.0	31.8	8.0	23.8	19.2	4.6	16.5
112	1	26.3	22.1	32.8	8.9	23.9	16.9	7.0	17.3
112	2	26.9	20.6	28.3	6.3	22.0	16.7	5.4	19.6
112	3	27.2	19.8	30.9	7.6	23.3	18.7	4.6	17.4
				Stem					
56	1	17.3	13.4	63.2	18.1	45.2	38.3	6.9	9.7
56	2	17.1	11.3	60.1	35.1	25.1	19.4	5.7	10.1
56	3	21.1	10.3	57.6	11.5	46.1	38.5	7.6	10.3
56	4	23.6	9.8	57.9	10.5	47.4	41.0	6.4	10.5
56	5	22.4	9.44	58.5	9.3	49.2	42.5	6.7	9.7
56	6	24.0	10.0	56.7	9.8	46.9	39.8	7.1	9.4
84	1	13.8	8.9	68.9	14.5	54.3	44.2	10.1	6.9
84	2	22.3	8.6	62.7	12.0	50.7	41.5	9.2	8.2
84	3	25.7	7.0	65.6	11.0	54.6	43.7	10.9	6.8
84	4	32.6	6.5	66.4	13.3	53.2	43.5	9.7	6.6
112	1	27.3	6.9	70.4	19.3	51.1	41.9	9.2	5.2
112	2	28.2	6.9	69.2	14.9	54.3	43.7	10.6	6.0
112	3	33.9	5.2	69.6	16.8	52.8	42.4	10.4	5.2

Quality Variations and other Forage Comparisons

Planting density produced significant variations on leaf CP and stem DM, hemicellulose and total ash. However, differences were slight from the biological point of view. The rest of the components showed very narrow variation from the mean. In general, spacing did not influence the nutrients present in leaf and stem. Cutting height statistically also affected leaf DM, CP, cell wall, ADF and total ash, but only stem DM. These differences were less than 1 percent, which are not biologically important to favour a certain cutting height.

Contrary to what had been found for planting density and cutting height, cutting frequency had a significant effect on nutrient accumulation, except for leaf hemicellulose and lignin. At frequent cuttings, leaf and stem CP and leaf cell wall were high. At longer cutting, stem NDF and all cell wall components increased. Total ash in leaves increased with cutting interval, whereas the opposite occurred with the stem. Chemical composition was very different between leaf and stem, except for the percentage of DM. Leaf CP was three times that of the stem, and leaf cell wall and ADF were half. Cellulose was 75 percent of ADF in leaf and 80 percent in stem. Lignin was 25 percent of ADF in the leaf and 20 percent in the stem.

Mulberry had two to three times more ash in leaves than in stem. Piccioni reports, on DM basis, mulberry leaf CP of 16 percent, ether extract 4.1 percent, cellulose 6.9 percent and ash 11 percent. Rodríguez, Arias and Quiñones found CP concentrations of 21 percent without fertiliser and 24 percent with 80 kg of N/ha after each 6-week harvest. Increasing cutting frequency to 12 weeks, leaf CP decreased to 18 percent without fertiliser and to 16 percent with fertiliser. Eswara and Reddy reported DM and CP values of 30 percent and 14.5 percent respectively.

When comparing mulberry with tropical and subtropical grasses, CP is higher than in alfalfa (17 percent), young Orchard grass (15 percent), Pangola grass (1 percent) and Elephant grass (9 percent). Mulberry leaf has 33 percent NDF and the stem 64 percent, whereas alfalfa has 40 percent, Orchard grass 55 percent, Pangola grass 70 percent and Elephant grass 72 percent, Mulberry stem at 112 days has a cell wall equal to or smaller than any tropical grass. Lignin in subtropical grasses varies between 4-8 percent and in tropical grasses between 7 and 8 percent, whereas mulberry has 5.5 percent in the leaves and 8 percent in the stem.

Table 6. Hemicellulose, cellulose and lignin in the neutral detergent fibre (NDF) or cell wall of mulberry leaf and stem

Factor	NDF (% DM)	Hemi. (% of NDF)	Cellulose (% of NDF)	Lignin (% of NDF)	Hem. Cellulose
Spacing (cm)					
60	33.5	27.3	55.2	17.5	0.50
90	32.4	25.5	58.9	15.6	0.43
120	32.6	26.3	57.4	16.5	0.46
Height (cm)					
30	33.5	28.0	56.2	15.9	0.50
60	32.1	24.6	58.0	17.3	0.42
Frequency (d)					
56	34.3	26.6	56.9	16.6	0.47
84	32.3	27.2	57.8	15.2	0.47
112	30.7	24.8	56.7	18,5	0.44
Spacing (cm)					
60	63.2	22.9	63.3	13.7	0.36
90	63.2	22.9	64.0	13.1	0.36
120	64.4	25.3	61.5	13.3	0.41
Height (cm)					
30	63.6	3.2	63.2	13.5	0.37
60	63.6	24.2	62.6	13.2	0.39
Frequency (d)					
56	59.0	26.6	73.4	11.4	0.36
84	65.9	19.3	80.7	15.1	0.24
112	69.8	24.4	75.6	14.5	0.32

It can be deducted that mulberry leaf is superior to any subtropical forage and that mulberry stem compares to tropical grasses. Considering that net energy for lactation in cattle is calculated from TDN from lignin content in NDF or ADF, the high digestibility of mulberry is confirmed. Table 6 presents hemicellulose, cellulose and NDF for leaf and stem. Lignin represents between 15 and 18 percent of leaf cell wall and between 11 and 15 percent of stem cell wall, which is used to estimate the high digestibility of both fractions.

The main limitation of the use of mulberry stem as feed is not related to chemical quality per se but to the physical form in which it is offered to animals for their use. Many tropical feeds, rspecially in dry periods, are inferior to mulberry stem in terms of quality. Mulberry leaf, properly dried and ground, is of excellent quality and its characteristics make it most suitable for inclusion in compounded feed for highly producing dairy cattle.

Mulberry Leaves as Feeds for Ruminants

For example ruminant livestock in most areas of the United Republic of Tanzania are managed under semipastoral and agro-pastoral systems, feeding mostly on natural pasture and crop residues. The amount of high-quality pasture is usually sufficient during the rainy season but, as maturity advances, the nutritive value decreases. Therefore, available feed resources during the dry season are usually unable to provide sufficient nutrients for reasonable livestock productivity, and livestock generally lose weight, become susceptible to diseases and have reduced breeding performance.

Although supplementation using commercial feeds or by-products such as maize bran and oilseed cakes is widely used to improve growth rates and milk yields, these supplements are expensive and not easily available in remote villages. Similarly, improvement of low-quality roughage by physical and chemical treatment is expensive, difficult to apply in rural areas and may result in environmental pollution. In recent years there has been a growing interest in many tropical regions to identify potentially important feed sources among shrubs and trees for inclusion in ruminant diets.

The increased importance of trees and shrubs as a non-conventional feed resource has been recognised as one of the most effective means of improving forage supply in smallholder livestock production. Emphasis on the use of trees in farming systems is based on their high-quality leaves and pods and the overall role they play in the natural ecosystems and human and animal welfare improvement. There are a number of introduced and native trees that are important sources of nutrients for livestock in Tanzania. However, some of the introduced trees have had difficulties adapting to the local environment (e.g. *Leucaena leucocephala*).

Moreover, most native deciduous browse species shed their leaves during the dry season, and during this period most evergreen species are known to contain some physical structures or chemical compounds that defend them against herbivores. Efforts to explore the potential use of mulberry (*M. alba*) trees as livestock feeds in the semi-arid areas of central Tanzania were initiated by Shayo. There are positive indications that the tree could be an alternative for improving farm animal diets and reducing feed costs in many production systems. This has prompted more studies on the effect of mulberry leaves on growth and milk production in farm animals (e.g. goats), currently being carried out at the Livestock Production Research Institute in central Tanzania and at the Sokoine University of Agriculture.

Mulberry is both a temperate and a subtropical plant grown in many regions of the world, predominantly in eastern, southern and southeastern Asia, southern Europe, southern North America, northwestern South America and parts of Africa. In Africa, it has been reported to occur in humid, subhumid and semi-arid areas at an altitude up to more than 1 000 m above sea level. It is not well known when mulberry was introduced in Tanzania but it appears to have been in the country for many decades.

In the humid areas of the northern highlands mulberry trees are incorporated into the intensive farming systems and are widely used to feed sheep and goats in a cut-and-carry system. In some situations mulberry is used as live fence. However, there have been reports that some missionaries have established mulberry trees for wine making, using the fruit. Apart from the highlands of the north, mulberry is also present in the semi-arid areas of central Tanzania, in the coastal belt and in the southern highlands. So far, there have been no formal studies on the distribution and use of mulberry in the country.

However, survey studies in some villages in two districts of the semi-arid areas of central Tanzania showed that mulberry trees were found in 5 percent of homesteads. The trees were grown close to the dwellings and the number of trees per household did not exceed five. Some trees were found away from the homesteads, where they were left by farmers who previously occupied the land. The trees were used for a range of purposes: for fruit, shade, vegetables, medicinal purposes and as fuel wood. The leaves were rarely used for livestock feeding, since most farmers were not aware of this use.

There are several natural and management factors affecting the productivity of mulberry, such as soil and climatic conditions, mode of propagation and harvesting techniques. In general, longer intervals between defoliation have increased total yield; however, the proportion of inedible wood may also increase, leading to a decline in forage quality. On the other hand, as plant spacing is reduced, yield per plant decreases owing to competition, but total forage yield per unit area increases, as does the leaf:wood ratio. This was confirmed by Shayo in a trial to study biomass production at different spacing and harvesting intervals in the semi-arid areas of central Tanzania (Table 7).

Table 7. Mean yield of mulberry, by fractions, under different spacings at various periods of the year

Harvest (years)	*Spacing*	*Yield/plant (kg DM)*				*Yield/ha (tonnes/DM)*			
		Leaf	*Stem*	*Bark*	*Total*	*Leaf*	*Stem*	*Bark*	*Total*
1st-2	S1	0.59	0.97	0.18	1.74	16.9	28.2	5.3	50.4
	S2	0.65	1.37	0.28	2.30	3.4	7.1	1.4	11.8
	S3	0.63	1.12	0.22	1.97	8.5	15.1	3.0	26.6
1st-2	S1	0.08	1.65	0.46	2.19	2.3	47.8	13.2	63.3
+ 190 days	S2	0.31	2.14	0.54	2.98	1.6	11.0	2.8	15.4
	S3	0.14	2.00	0.49	2.64	2.0	27.1	6.6	35.7
2nd-120 days	S1	0.14	0.12	0.05	0.31	4.0	3.5	1.4	9.00
	S2	0.13	0.14	0.06	0.33	0.7	0.7	0.3	1.7
	S3	0.11	0.14	0.06	0.30	1.5	1.9	0.8	4.1
2nd-190 days	S1	0.05	0.16	0.02	0.22	1.3	4.5	0.6	6.4
	S2	0.06	0.17	0.02	0.25	0.3	0.9	0.1	1.3
	S3	0.05	0.17	0.02	0.23	0.6	2.3	0.3	3.1

Although yield per season increased with harvesting interval, the amounts and nutritive value of edible components decreased considerably due to senescence and drop of leaves. Tikader *et al.*, found that maximum mulberry leaf yields could be obtained by harvesting three times per year. The results in Table 7 show that mulberry harvested twice, at the end of rainy season and in the middle of the dry season, gave considerably higher amounts of fodder.

It is interesting to note that mulberry harvested during the peak of the dry season developed considerable numbers of shoots. As reported by Walker, regeneration of new shoots after pruning is independent of rainfall. The advantage with the new shoots is that they are more tender and nutritious than ordinary branches and grow more rapidly. This is also reflected by higher proportion of leaf:stem in younger plants and regrowths than older plants (Table 8).

Table 8. Proportion of mulberry fractions as affected by spacings and harvest time

Type	*Treatment*	*Part of the plant (%)*		
		Leaf	*Stem*	*Bark*
First harvest at 2 y	S1	33.6	55.9	10.5
	S2	28.3	59.6	12.2
	S3	31.8	56.8	11.4
	Mean	31.2	57.4	11.4
First harvest at 2 y + 190 days	S1	3.6	75.5	20.9
	S2	10.2	71.7	18.1
	S3	5.5	75.9	18.6
	Mean	6.4	74.4	19.2
Regrowth at 120 days				
	S1	44.7	39.4	15.9
	S2	39.3	42.5	18.2
	S3	36.1	45.4	18.4
	Mean	40.0	42.4	17.5
Regrowth at 190 days	S1	20.5	70.1	9.3
	S2	23.7	66.8	9.5
	S3	20.3	71.9	8.0
	Mean	21.5	69.6	8.9

Mulberry leaves are highly digestible and contain high concentrations of crude protein (CP) and minerals, and low cell wall contents (Table 9). This suggests that mulberry could be used to minimise nutrient deficiencies faced by grazing animals in the semi-arid areas of central Tanzania. Furthermore, the bark is reasonably digestible, with CP concentrations higher than in dry season pastures and crop residues. Older leaves contain lower concentrations of CP than young leaves (Table 9). Young leaves and bark are more degradable than old leaves and bark.

Table 9. Chemical composition and digestibility of mulberry leaves and bark in central Tanzania

Parameter (% DM)	*Part of the plant*			
	Leaf	*Bark*	*Leaf*	*Bark*
Ash 14.3	6.1	13.3	6.3	
Crude protein	18.6	7.8	14	8.7
Neutral detergent fibre	24.6	46.8	27.6	44.5
Acid detergent fibre	20.8	36.9	25.1	38.3
Hemicellulose	3.8	9.9	2.5	6.2
Acid detergent lignin	8.1	7.1	-	-
Cellulose	12.6	29.7	-	-
Acid detergent insoluble ash	2.5	1.1	-	-
In vitro DM digestibility	82	60	-	-
In vitro organic matter digestibility	-	-	89	85

The rate of in sacco degradation of leaves was high, 70-80 percent at 24 hours of incubation. The study also showed that degradability of the old bark was maximal after 72 hours. Another advantage of mulberry leaves is that they contain considerably lower levels of total phenolics compared with leaves and pods of most native browse trees and shrubs in central Tanzania, such as Faidherbia albida, Acacia tortilis, Acacia nilotica, Delonix elata and Dichrostachys cinerea. Phenolic compounds have a capacity to bind to carbohydrates, plant proteins, salivary mucoprotein and gastrointestinal enzymes, thereby reducing protein and cell wall digestion. Makkar *et al.* classified mulberry as low tannin fodder tree.

6

Mulberry Leaf Supplement for Sheep Feed

The use of crop residues as basal diets for fattening cattle and lambs has been promoted in China during the last decade with much success. With growing lambs, Liu *et al.* observed that the growth rate was dramatically increased when the ammonia bicarbonate rice straw (ABRS) diet was supplemented with small amounts of rapeseed meal (RSM), and that the benefits of ammoniation were least when a high level of RSM was used. Sericulture, based on mulberry (*M. alba*) leaves, is an important farming activity in China, with over 106 ha planted in China and 105 ha in Zhejiang Province. Yield of fresh leaves is in the order of 15 to 22 tonnes/ha/year.

Mulberry leaves are relished by sheep and goats and have a high nutritive value with a protein content of about 20 percent of DM. Roothaert observed that dairy heifers had higher voluntary intake, and thus higher potential of milk production, when consuming mulberry fodder rather than the cassava tree (*Manihot glaziovii*) and Leucaena (*Leucaena diversifolia*). Mulberry leaves could be considered an appropriate supplement for sheep fed a basal diet of ammoniated straw, replacing partially or totally the oilseed meals, which could then be used in monogastric diets.

Some researchers done a study on mulberry plants to evaluate the nutritional value of mulberry leaves collected from clones at different stages of maturity. And they also determine the effect of mulberry leaves as a supplement, substituting RSM, for growing lambs fed ABRS. The results of the studies are briefly described below.

Evaluation of the Nutritional Value of Mulberry Leaves

Sampling of Mulberry Leaves

Clones used were Tuantou Heyebai (TH), Husang No. 9 (HS), Tongxiangqing (TX) and Nongsang No. 8 (NS). Leaves were sampled six times in 1998, three in the spring (28 April, 14 and 29 May); and three in autumn (28 August, 30 September and 30 October). Samples were taken in the morning, weighed immediately and ovendried at 65°C. Subsamples were milled to pass a 2 mm -sieve for further nutritional evaluation. All samples were analysed for CP and true protein (TP) following AOAC; for NDF. Amino acid (AA) contents were determined using an AA analyser.

The nutritional value of mulberry leaves was evaluated with *in vitro* gas production (GP) of Menke and Steingass with calibrated glass syringes. Samples for GP determination were ground with a hammer mill to pass a 1-mm-screen. About 200 mg (DM) samples were introduced into syringes with rumen liquor collected from two rumen fistulated Huzhou sheep, fed an ammoniated rice straw diet. The GP data were then fitted to the equation GP = a+b (1-exp(-ct)), where a, b and c are constants and GP is the gas production from the substrate at time *t*.

Results for each season were analysed according to a 3 x 4 factorial design. The difference of means was tested using Duncan's new multiple range test. The ABRS was prepared by the "stack method". Fertiliser grade ammonium bicarbonate was used and the weight proportion of straw:bicarbonate:water was 100:10:30. The straw was treated under ambient temperature for 30 days. The treated straw was then exposed to the air for a maximum of 24 hours before feeding to allow free ammonia to escape. Mulberry leaves were collected at Tongxiang Silkwork Breeding Farm. All leaves were harvested in autumn and airdried before storage.

Forty-five growing Huzhou lambs with an initial weight of 16-18 kg, were divided into five equal groups, according to sex and weight, and randomly allocated the following treatments:

A: 100 g RSM

B: 75 g RSM + 60 g mulberry leaves

C: 50 g RSM + 120 g mulberry leaves

D: 25 g RSM + 180 g mulberry leaves

E: 240 g mulberry leaves

Gupplementation with RSM at 100 g/day was based on the result in a previous study. All five supplementary treatments were calculated to have similar CP content. All animals had ABRS, mineralised salt block and water ad libitum and 100 g/head/day of ground maize.

In vitro Gas Production Test

GP was measured to compare the nutritional value of different diets. In order to analyse the associate effect on GP parameters, only the mixtures of RSM and mulberry leaves at different ratios were also incubated. The proportions of both supplements for different combinations were based on their ratios in the corresponding diets in feeding trials. The results were analysed by one-way analysis of variance. The difference of means for the five treatments was tested by using Duncan's new multiple range test.

Chemical Composition of Mulberry Leaves

The chemical composition of mulberry leaves is shown in Table 1. There were no significant differences (P >0.05) among mulberry clones in DM, CP, TP and NDF (for spring leaves). Only the clone Tuantou Heyebai had lower NDF (P <0.05) in the autumn. Sampling time had a great effect on chemical compositions. For spring leaves, contents of CP and TP were slightly higher at mid stage than at early or late stage, whereas the CP content of autumn leaves decreased significantly (P <0.05) with maturation. The NDF content increased with maturation (P <0.05) in both seasons.

Little seasonal differences were found in the contents of CP and TP of mulberry leaves. Average CP contents were 21.1 and 20.9 percent of DM, and the TP accounted for 88.2 and 85.8 percent of CP in spring and autumn, respectively. However, the NDF content was lower for mulberry leaves in spring (38.8 percent DM) than for those in autumn (41.4 percent). Except for a few amino acids (AA), no significant difference (P >0.05) was observed in individual AA content among clones (data not included). There were little differences in total, essential or non-essential AA among clones. AA content tended to increase with the time, but the differences showed no statistical significance.

Table 1. Chemical composition of mulberry leaves in spring and autumn

Sample	*Dry matter (%)*	*Crude protein (% DM)*	*True protein (% CP)*	*NDF (% DM)*
Spring leaves				
Date				
28 April 98	25.1	21.1	87.5	34.6
14 May 98	25.2	21.9	90.4	38.9
29 May 98	26.6	20.0	86.3	42.9
Clone				
Tuantou Heyebai	24.4	21.6	86.7	39.5
Husang No. 9	24.1	20.9	89.6	37.5
Tongxiangqing	23.7	20.9	88.3	39.7
Nongsang No. 8	23.6	20.8	88.5	38.5
SE	0.10	0.17	1.16	0.59
Autumn leaves				
Date				
28 August 98	25.9	22.3	86.9	36.7
30 September 98	29.9	21.4	85.7	40.4
30 October 98	33.8	18.9	84.7	47.2
Clone				
Tuantou Heyebai	30.4	21.9	84.4	38.9
Husang No. 9	29.7	20.3	86.4	40.8
Tongxiangqing	29.6	19.6	87.5	42.6
Nongsang No. 8	29.8	21.7	84.9	43.4
SE 0.42	0.23	0.57	0.46	

Nutritional Value of Mulberry Leaves

The nutritional value of mulberry leaves based on the GP test is presented in Table 2. In spring, clone Nongsang No 8 had higher nutritive value ($P < 0.05$). There was little difference between clones in the GP24 and potential GP for autumn leaves, though rate of GP was slightly higher for clones Nongsang and Tongxiangqing. The estimated organic matter digestibility (OMD) showed a similar tendency to GP parameters. Similar to CP content (Table 1), mid-spring samples tended to have a higher GP ($P < 0.05$) than early or late sampling, while for autumn leaves the GP in late season was lower ($P < 0.05$), with little difference between those at early and middle season.

Table 2. In vitro gas production (GP) parameters and estimated organic matter digestibility (OMD) of mulberry leaves

Sample	*GP at 24 hours(ml)*	*Potential GP (ml)*	*Rate of GP (% h-1)*	*OMD(%)*
Spring leaves				
Date				
28 April 98	43.7	47.8a	9.70	69.2
14 May 98	46.9	52.2	8.79	71.3
29 May 98	38.5	43.3	9.25	65.6
Clone				
Tuantou Heyebai	41.9	45.7a	7.26	69.1
Husang No. 9	43.1	47.4a	9.02	68.8
Tongxiangqing	39.6	43.9	8.90	65.9
Nongsang No. 8	47.6	52.7	9.50	71.9
SE 0.94	1.04	0.11	0.74	
Autumn leaves				
Date				
28 August 98	32.5	38.7	7.15	61.4
30 September 98	31.7	38.3	6.69	60.3
30 October 98	28.7	35.4	6.18	56.3
Clone				
Tuantou Heyebai	31.0	37.8	6.57	60.0
Husang No. 9	30.8	38.0	6.16	58.8
Tongxiangqing	30.8	36.7	7.01	58.4
Nongsang No. 8	31.3	37.4	6.95	60.1
SE	0.55	0.60	0.11	0.47

In general, the nutritional value of spring leaves was much higher than autumn leaves. The spring mulberry leaves with OMD of 65.6-71.3 percent are comparable with some leguminous hays such as alfalfa and vetch. According to farmers' practice, the twigs of mulberry trees must be cut and modified in late May or early June in order assure to ensure leaves for the silkworm. These twigs and leaves may be dried and stored for winter use.

Dry Matter Intake

CP and NDF contents for mulberry leaves used in feeding trials were 23.0 and 43.7 percent DM, and the corresponding values for RSM were 42.3 and 51.1 percent DM, and for ABRS, 13.3 and 63.9 percent DM, respectively. The rumen degradation results are shown in Table 3. The rates of disappearance of DM were higher for mulberry leaves than for RSM, but those of CP were lower. CP fractions degrading rapidly (a) or slowly (b) were much lower for mulberry leaves than for RSM. Rumen escape protein was higher in mulberry leaves than in RSM.

Table 3. Constants of DM and crude protein of the equation p=a+b(1-exp(-ct)) together with 48 hour rumen degradability (D_{48}) of rapeseed meal and mulberry leaves

	Rapeseed meal		*Mulberry leaves*	
	DM	*CP*	*DM*	*CP*
a (%)	19.2	32.6	20.5	19.5
b (%)	47.9	67.4	53.9	50.2
c (%/hour)	2.70	2.07	3.10	2.57
se	2.07	2.50	0.65	0.84
a+b (%)	67.1	86.5	74.4	69.7
ED with passage rate at				
2.00 percent/hour	46.7	66.9	53.3	47.7
4.00 %/hour	38.5	55.6	44.0	39.1
D_{48} (%)	54.5	76.2	62.1	54.9

Animals consumed all rapeseed meal, ground maize and mulberry leaves offered. The intake of ABRS was slightly increased when supplementing with the mulberry leaves, and hence the total intake increased with the increase in mulberry leaves (Table 4). Total DM intakes were 451, 455, 495, 540 and 590 g/day for lambs on diets A, B, C, D and E, respectively. Basal ammoniated straw accounted for more than 50 percent of total diets in all treatments, and intake from forage exceeded 85 percent of total diets when the RSM was fully substituted by mulberry leaves.

Table 4. Dry matter intake and growth performance of lambs

Parameter	*Diets (g Mulberry: g RSM)*				
	0:100	*60:75*	*120:50*	*180:25*	*240:0*
N° of animals	9	9	9	9	9
Feed intake (g DM/day)					
Rapeseed meal	88	66	44	22	0
Mulberry leaves	0	52.30	104.50	156.60	209.00
Ground corn	86	86	86	86	86
Ammoniated rice straw	277.10	250.50	260.60	275.80	295.60
Total intake	451.10	454.80	495.10	540.40	590.60
Initial weight (kg)	18.00	17.70	17.80	16.20	18.40
Growth rate (g/day)	58±4	47±8	40±6	46±5	55±3
Feed efficiency (kg/kg)					
Total intake/gain	7.78	9.67	12.37	11.75	10.74
Concentrate/gain	3.00	3.23	3.25	2.35	1.56
Feed cost/kg gain	4.56	5.39	6.25	5.38	4.47

Intake of the basal diet is usually decreased because of a substitution effect when a straw-based diet is supplemented with forage. Tharmaraj *et al.* observed a decline in the DM intake of both ammoniated and untreated RS when supplemented with *Gliricidia*, a leguminous tree. Liu *et al.* found that inclusion of milk vetch silage at a level higher than 23 percent of diets reduced the intake of ammoniated rice straw by growing heifers. In the present study the mulberry leaves accounted for 11, 21, 29 and 35 percent of dietary DM intake in diets B, C, D and E, but the ABRS intake was even slightly increased with the rising amounts of mulberry leaves.

Lamb Growth

Lamb growth rates are shown in Table 4. The growth rate in diets with only RSM was comparable to that obtained in a previous trial by Liu *et al.*, who supplemented ABRS diet with 100g RSM and 100g rice bran. The animals supplemented with mulberry leaves grew the same as those with RSM, but the growth rates were lower when both supplements were given together. There was no difference among sex and groupings. In the limited literature on mulberry leaves as animal feeds, Leng mentioned that dairy cows achieved up to 18 litres of milk/day on forage supplemented with fresh mulberry foliage. In his perspective, the production rate on high intakes of tree foliages such as mulberry may be as good as those of cattle on ammoniated straw and supplemented with 1-1.5 kg/day of cottonseed meal.

While feed efficiency was higher when receiving only RSM, concentrate consumption per kg of weight gain was lower when a higher level of mulberry leaves was supplemented. Compared with other treatments, feed cost per kg gain was lower in the diets with only RSM or mulberry, the lowest in the latter. When mulberry leaves were used as the supplement to substitute for RSM, there was a benefit of 0.09 yuan/day, equivalent to yuan 9.00 for the usual fattening period of 100 days.

The results of *in vitro* GP tests for different diets are summarised in Table 5. The GP for diets with low levels of mulberry was lower than for other diets. The potential GP was significantly higher in diet with only RSM. The GP48 value showed a similar trend to the potential GP, although the difference between diets was not statistically significant. These results suggest a negative associative effect between rapeseed and mulberry leaves.

Table 5. Parameters of in vitro gas production (GP) for different experimental diets

Parameters	*Diets (g Mulberry: g RSM)* 0:100	60:75	120:50	180:25	240:0	*SE*
a (ml)	2.5	3.0	2.0	1.7	2.9	0.3
b (ml)	36.4	33.3	32.9	33.6	34.4	3.6
a+b (ml)	38.9	36.3	34.5	35.6	37.7	1.4
Rate of GP,%/hour	3.56	4.01	4.38	5.33	4.43	0.07
GP_{48} (ml)	31.2	30.0	29.4	31.5	31.5	4.5

Table 6 presents the result of GP tests when only mixtures of mulberry leaves and RSM were incubated. The potential GP and rate of GP for mulberry leaves (PE) were higher than for RSM (PA), but the GP parameters were not improved proportionally when mulberry leaves were increased.

Table 6. Parameters of in vitro gas production of mixtures of mulberry leaves and rapeseed meal

Treatment	*PA*	*PB*	*PC*	*PD*	*PE*
Mulberry:RSM	0:100	41:59	68:32	86:14	100:0
GP parameters					
a (ml)	3.3	2.9	2.7	1.1	-0.2
b (ml)	29.1	29.5	32.4	38.0	41.7
a+b (ml)	32.4	32.4	35.1	40.1	41.5
c (%/hour)	4.22	4.41	5.92	6.15	6.10
GP_{48} (ml)	27.7	28.3	32.7	37.3	39.2

There was apparently a negative associative effect between mulberry leaves and RSM where the estimated GP values were much lower than the measured ones. This may partially account for the growth rate of lambs offered different diets (Table 4). There may be some secondary plant compounds in mulberry leaves that exert a detrimental effect on ruminal microbes or reduce the utilisation of the dietary nutrients. Further study is needed to clarify these aspects.

7

Mulberry Genetic Resources

India is recognised as one of the twelve-mega biodiversity centres of the world and harbouring 11.9% of the world flora. Mulberry belongs to the fig family Moraceae under the order Urticales that is considered to be advanced in the woody flowering plants. Mulberry (Genus Morus L.) with high biomass production and protein rich foliage not only nurturing the entire sericulture industry in India besides, used extensively in agroforestry and horticulture programmes. India has attained the status of second largest silk producer in the world with galloping strides of raw silk production (15,000 metric tonnes accounting 16.1%) after China.

In India, efforts have been made to use primary gene pools constituted predominantly cultivars belong to Morus indica and Morus alba to tailor the desired varieties. However, the promising genes exist in the secondary and tertiary gene pool in the wild species of Morus laevigata and Morus serrata and other wild species can be utilised for gene transfer by applying biotechnological and molecular techniques. Realising the importance of mulberry germplasm for sustainable development of sericulture in India, Central Sericulture Germplasm Resource Centre (CSGRC) was established in

India which is presently holding 908 diverse mulberry germplasm accessions representing 26 countries belong to 13 mulberry species in the ex situ field gene bank. Characterisation and evaluation of 628 mulberry accessions has been completed and the trait specific promising accessions have been identified. The germplasm information is documented in the Mulberry Germplasm Information System (MGIS) with query-based retrieval.

Realising the importance of impending global climatic changes and threatened sustainability of biodiversity wealth in India at faster rate, systematic survey and exploration for collection of mulberry biodiversity have gained greater momentum in the recent past. Consequently, CSGRC at Hosur has been aptly established in the year 1990 by Central Silk Board (CSB), Ministry of Textiles, and Government of India under prestigious National Sericulture Project. It is the nodal agency for mulberry germplasm management in India and recognised as National Active Germplasm Site (NAGS) for mulberry by National Bureau of Plant Genetic Resources (NBPGR), New Delhi, India under Indian National Plant Genetic resources System (IN PGRS).

CSGRC, Hosur so far conducted 46 survey and exploration trips covering more than 50 districts from Himalayan belt to Andaman and Nicobar Islands covering forest areas, biosphere reserves, national parks, back yards, kitchen gardens, agricultural lands and farmhouses etc. and collected 516 diverse mulberry germplasm resources which mainly belong to 4 Indian species. Out of 4 species of genus Morus reported in India, *M.laevigata* is distributed throughout India both under natural and managed habitats and *M. serrata* is confined to northwest India in wild condition. *M. indica* and *M. alba* are mostly available in cultivated forms.

The sericulture research institutes like Central Sericultural Research and Training Institute (CSRTI) at Berhampore (West Bengal), Mysore (Karnataka) and Pampore (Jammu and Kashmir) have started collection of Morus species in Central Himalayas, North-Eastern India, North-Western Himalayas, Kumaon and North Eastern India and Western ghats and Kerala. All these germplasm resources are collected and introduced in the field gene bank.

Conservation Strategies

India being signatory to both CBD and World Trade Organisation (WTO) needs to develop the efficient conservation strategies of its vast seribiodiversity best suited to India's national interest. Exploration for collection of mulberry biodiversity and development of efficient conservation strategies has gained momentum in the recent past in India. The conservation of mulberry genetic resources includes their augmentation, safe holding for medium and long-term preservation; protection in natural habitats and it is interlinked to the sustainable utilisation of germplasm to justify long-term investments on managements and maintenance of the germplasm.

Studying the geographical spread, distribution and genetic architecture, arborescent nature and physiological storage behaviour of mulberry, two basic conservation strategies i.e. *Ex situ* and *In situ* methods composed of various techniques covering entire gamut of genetic diversity have been worked.

Ex situ Conservation

Mulberry being perennial outbreeding tree exhibits high degree of heterozygosity. Hence, for conservation of mulberry outside its natural habitat, *ex situ* field gene bank (clonal repository) has been developed by planting vegetative clones of mulberry accessions for maintaining the genetic integrity of the conserved material. Rooted cuttings (saplings) after six-month establishment in the nursery are transplanted in the field gene bank. Saplings of some exotic accessions and wild Morus species with poor rooting rate, are developed through bud grafting to the local scions and then established in the base collection.

The plants (four plants/accessions) are maintained as a dwarf tree with spacing of 2.4 ´ 2.4 m between plants with the crown height of 1.5 m following recommended cultural practices and plant protection measures. Pruning is being followed once in a year (June-July) to renew the germplasm with new sprouts. However, the species, which cannot sustain repeated pruning are left unpruned. The entire collections in field gene bank are fenced and protected. Presently, the ex situ field gene bank of CSGRC, Hosur holds 908 mulberry accessions (Indigenous-647 and Exotic-261) from diverse genetic and geographical origin representing 13 Morus species collected from 26 countries.

CSGRC, Hosur being the nodal agency for mulberry germplasm in India maintain the entire mulberry germplasm available in the country. However, the Sericulture Research Institutes, Universities, State Sericulture Departments are also maintain the mulberry germplasm for research purpose, which also serves as backup conservation centres. Each accession contains permanent label with unique identification number. National Accession Numbers have been provided for all the mulberry accessions conserved in the *ex situ* field gene bank by NBPGR, New Delhi for protection of mulberry genetic resources at global level.

On-farm Conservation

The advent of high yielding varieties of mulberry like V-1 and S-1635 and their large scale spread in the traditional sericultural zones under monocropping pattern replacing the local landraces and more particularly in the irrigated system which leads to reduced genetic base and increased the genetic vulnerability of the crop. The seribiodiversity otherwise greatly threatened because of unlawful habitat destruction, natural calamities, fragmentation of forests and social disruption and this large scale genetic wipeout disturb the coexistence of sericigenous flora and fauna. Under these circumstances on-farm conservation linked with Farmers Participatory Breeding (FPB) to be given due emphasis for achieving twin goals of sustainable conservation of biodiversity and in turn utilisation on-farm biodiversity.

In India, rich Morus diversity exists under managed habitats i.e. in the backyards, kitchen gardens, farmhouses, horticultural gardens, agricultural lands and roadside plantations. These are the first hand selections of the farmers and tribals for varied utilisations hence; conservation of potentially interesting alleles and development of diversity is promoted. In mulberry the wild species like *Morus laevigata* and *Morus serrata* and other wild species, which do not get attention in the formal sector for cultivation for sericulture purposes. So, these associated valuable species or otherwise utilised for non-sericultural purposes (horticulture and agroforestry) flourish well in the on-farm conservation procedures promoting farmers/tribals livelihood development while conserving Morus biodiversity. This particular sector of Morus biodiversity lies mainly in the public domain, which needs to be taken care and bring them under definite set of legal framework of Indian Forest Act.

Cryopreservation Technology

Conservation of mulberry germplasm in the field gene bank is simple and technically less demanding. However, maintaining a large collection is costly, requiring huge resources. Besides, it is risky since it is exposed to different biotic and abiotic stresses. Cryopreservation technology forms an alternative to *ex situ* conservation strategy for crop like mulberry, which is vegetatively propagated utilising liquid nitrogen (LN) at ultra low temperature of -150°C (vapour phase). Cryopreservation method ensures the genetic stability of the germplasm and provides an alternate cost effective non-dependence on electricity long-term strategy for conservation of mulberry germplasm.

Keeping this in view, establishment of *in vitro* and cryopreservation laboratory facilities at CSGRC, Hosur is planed in collaboration with NBPGR, New Delhi. Efforts have been made to standardise the efficient techniques on cryopreservation of winter buds, embryonic axes, pollen of mulberry accessions comprising different species, landraces, local cultivars, wild and polyploid accessions in liquid nitrogen (-196°C). *In vitro* conservation techniques involve conservation of active collections under active growth stage and base collection germplasm under suspended growth stage using growth retardants.

Even though lot of work on mulberry regeneration with or without callus formation in *in vitro* has been reported. Very little work has been carried out in *in vitro* conservation of mulberry genetic resources. Single shoots of *M. nigra* L stored on multiplication medium at 4°C for 16-hour photoperiod survived only for six months. Survival was enhanced to 42% at nine months by storing them at 25°C with activated charcoal as supplement. High viability (80%) for six months was observed in 15 genotypes of *M. alba* stored at 4°C dark in shoot proliferation medium.

Rooting was observed in all the shoots and shoots retained their multiplication potential. Cryopreservation procedures have been standardised for about 100 different plant species cultured in various ways including cell suspension, callus, apices and zygotic and somatic embryos. Cryopreservation, possibility for the first time was demonstrated using mulberry twig by Sakai. Since then, considerable work on cryopreservation of mulberry has been undertaken especially in Japan. Shoot tips of pre-frozen winter buds of *M. bombycis* Koidz. was able to withstand storage in LN.

However, grafts and cuttings did not survive in LN. With modification of this method it was able to regenerate plants of *M.multicaulis* P. through tip culture of frozen winter buds. Shoot segments were pre-frozen at - 3°C for 10 days, - 5°C for three days, - 10°C for 1 day and - 20°C for one day before immersion in LN. Buds were cultured on MS medium after thawing in air at 0 to 20°C. Survival rate was 55 to 90%. Excised shoot tips from winter buds of M. bombycis Koidz. pre-frozen at 10°C per day. Prior to pre-freezing at -20°C partial dehydration to 38.5% improved the recovery rates. The survival rates of the winter buds stored in LN from the month to 3-5 years did not change.

Direct dehydration with silicon gel at 25°C of excised shoot tips (2 mm long) from winter bud could be done before immersion in LN. With decreasing water content shoot formation increased and at about 19% of water content, a maximum of 80% survival rate was observed. Encapsulation by alginate coating of winter hardened shoot tips of many Morus species had 81% of shoot formation with 22-25% water content. *In vitro* grown shoot tips of thirteen cultivars of mulberry were tested for cryopreservation. Slow freezing (0.5°C/min. to - 42°C), vitrification (PVS2, 90 min) and air- drying (24% water content) or encapsulation dehydration (33% water content) was tested for survival, which ranged from 40-81.3%.

It was also reported the long-term storage of mulberry winter buds by cryopreservation. Winter buds from *M. bombycis* with about 10 mm vascular tissue were kept at 0°C for 1 day before freezing. Buds were cooled to 10°C steps at daily intervals from 0 to 30°C prior to immersion in LN or before transferring to -135°C. After storage, buds were rapidly thawed at 37°C in a water bath and then cultured on MS medium supplemented with 2% fructose and 1 mg/l with 6-BAP. Rate of shoot formation did not vary much in buds stored in LN or deep freezed at -135°C after a storage period of 3-5 years.

In India, the application of *in vitro* technique for mulberry conservation has been recently attempted. *In vitro* technique has been attempted in mulberry mainly for propagation of popular mulberry varieties and poor rooting materials, evolution of some clonal variants and elite materials and screening of genotypes for tolerance to salt and somatic stresses. Bapat et al. developed method for propagation of *Morus indica* L. (Mulberry) by culturing encapsulated shoot buds. Conservation of mulberry germplasm accessions through cryopreservation technique has been recently attempted in India.

In situ Conservation

In situ conservation promotes the conservation of ecosystem and natural habitats and the maintenance and recovery of viable population of species, which can survive, and best perpetuate in their natural microclimate.

It also simultaneously permits continued evolutionary development under natural selection pressures, thereby promoting the fitness of the species. In situ conservation demands the establishment of nature or biosphere reserve and national parks to protect the endangered species. The National Committee on Environmental Planning and Coordination (NCEPL) and Man and Biosphere (UNESCO) already identified 14 Biosphere reserves in India and among them Uttarkhand, Nandadevi, Namdapha, Kaziranga, Manas, Nokrek, North Andaman and Great Nicobar are the potential reserves for *in situ* conservation of mulberry.

Keeping this in view, efforts have been made to collect information on the location of availability of mulberry germplasm with details on "declared protected area network of India" including biosphere reserves, national parks, wild life sanctuaries etc. Mulberry is not fully protected under Indian Forest Act in many parts of India except in some states of North-East India. Survey map, exact location, landowner with postal address, survey number and its jurisdiction are not available with CSB units or State department. Under the circumstances, a suggestive and advisory role can be contemplated with greater thrust on repeated survey and exploration.

Since ages, mulberry is being worshiped by tribals of Uttaranchal of Himalayan region and conserved in the name of "Sacred grooves". The sacred mulberry tree (*M. serrata*) at Joshimath (altitude of 2 000 above MSL) being worshipped by the pilgrims of Badrinath, is said to be the oldest (1200 years) and biggest mulberry tree (21.6 m of circumference) in the world. The great Indian sage Adiguru Shri Shankarcharya is said to have meditated under this tree. Because of this reason all *M. serrata* trees are found unaxed and worshiped with taboo at several places in Garhwal and Kumaon regions of Himalayas.

Similarly, trees of *M.laevigata* are conserved in Punjab. Attempts are being made to locate the concentrated areas of wild/natural mulberry trees of *Morus laevigata* at Island of Andaman & Nicobar biosphere, Arunachal Pradesh and areas of Himalayan region. Large numbers of naturally grown *M.laevigata* trees are also available in Eastern Himalayas of south and west Sikkim, West Bengal, Assam, Meghalaya and Manipur. Besides, developed gene pool of *M.alba, M.indica* and *M.laevigata* are also available in western ghats covering Kerala, Tamilnadu and Karnataka, Madhya Pradesh, Maharastra, Rajasthan, Yarcaud hills of Tamilnadu and Simlipal reserve forest of Orissa etc. which are adapted to local environment for many years are also required to be protected under in situ conservation.

Management and Utilisation

Characterisation of mulberry genetic resource determines the structural and functional attributes of the accessions and is highly essential for their genetic identity under IPR regime. The comprehensive information on variability recorded helps breeders and users for effective crop improvement. For choosing the donors for hybridisation programmes, the background information is highly essential. The data need to be standardised in terms of terminology, measurement codes, encoding methods and data recording procedures and hence, characterisation becomes mandatory and forms a bridge between conservation and utilisation. Systematic characterisation of mulberry germplasm resources was started after establishment of CSGRC, Hosur in year 1990.

Morphological Characterisation

Morphological characterisation is mainly based on visual

observation and wherever it is necessary it is supported by quantitative measurements. These characters are strongly heritable in nature and manifested equally in all the environments and hence used in the genetic identify of the accession and this also enable an easy and quick discrimination between the phenotypes. Thus, the characterisation and estimation of the morphological variability provides useful measure of genetic diversity prevalent in the gene pool. At CSGRC, Hosur morphological characterisation has been completed for 628 mulberry accessions.

Reproductive Characterisation

The knowledge of floral structure and reproductive behaviour of mulberry accessions are of utmost important to undertake any breeding programme for maximum utilisation of genetic resources for future needs. At CSGRC during the first phase 617 mulberry accessions (Indigenous-446 & Exotics- 171) have been characterised with 26 reproductive descriptors. Three major groups viz. dioeciously male, female and monoecious with regard to sex expression and sexual composition of various accessions are observed.

Out of 617 mulberry accessions studied 368 (59.64%) produced strictly pistillate (female) inflorescence, 75 (12.48%) staminate (male) inflorescence and 172 (27.88%) are monoecious types. Monoecious accessions with high leaf yield base may utilised for developing inbred lines. Increased degree of dioecy (femaleness) was observed in most of the locals and wild genotypes and this prevalence of unisexuality and preponderance of femaleness in mulberry irrespective of genetic and geographical origin might have resulted through evolutionary changes by suppression and progressive reduction of one sex in bisexual flower.

Leaf Histological Characterisation

Leaf is the vital organ in a plant where the major physiological activities take place. Largely stomatal characters possibly influence stomatal resistance, transpiration rate, and photosynthesis rate. The assimilatory tissue such as number of chloroplast in the guard cells of the stomata is directly correlated to the ploidy status. Physio-anatomical features related to drought tolerance in mulberry were reported. Thus, the leaf anatomical characterisation of mulberry is highly essential not only for the phylogenetic studies but also for the identification of stress tolerant varieties. At CSGRC during the first and second project periods 628 mulberry accessions (Indigenous-455 and Exotics-173) have been systematically characterised with fourteen anatomical descriptors.

Cytological Characterisation

Mulberry is characterised by different ploidy levels ranged from x to 22x. Among the different ploidy levels triploids are considered to be superior in most of the economical characters and wider adaptability to different agro climatic conditions. Most of the genotypes which belong to *M. alba* L. and *M. indica* L. are diploids (2n = 2x =28). The wild species of mulberry belong to *M. laevigata Wall.* and *M. serrata Roxb.* are available in different ploidy levels.

Hexaploid mulberry varieties belong to *M. serrata Roxb.* (6x=2n=84) *M. cathyana* has been reported. Octaploidy in *M. cathyana,* docosoploidy in *M. nigra* and Polysomaty in *M. multicaulis* and *M. alba* was also reported. The karyotype morphology will help in understanding the phylogenetic relationship among the different species and its distribution has a great value in modern taxonomy.

Molecular Characterisation

WTO requires its member countries to accord protection to plant varieties under TRIPS agreement. The protection intended to buoy innovations, which can be provided by patents, by an effective sui generis system or by any communication thereof. Molecular markers on the other hand exhibit high degree of non-tissue specific polymorphism and simple inheritance pattern with minimum influence of environmental and epistasis. In the last two decades, geneticist has started using biochemical and molecular techniques to add valuable information for characterisation of germplasm. In addition to the fact this genetic information can be generated quickly, they are environmentally independent, accurate and reliable.

Isozyme markers have been successfully used to characterise plant genetic resources. Even though, isozyme analysis is simple and relatively cost effective they are confine to small portion of genome and often they do not offer sufficient polymorphism to reveal the genetic difference. In the pursuit, molecular approaches like RFLP, RAPD, microsatellite markers, AFLP etc. effectively overcome this problem by offering high level of polymorphism. These techniques along with computer based software programmes are being routinely used in characterisation and germplasm related studies like marker based gene tagging, gene cloning, diversity analysis and marker assisted selection of desirable genotypes all over the world.

Analyses of isozyme

At CSGRC, Hosur the isozyme study was initiated in 17 representative Morus species using peroxidase enzyme system to assess the allelic variation at isozymic loci and to ascertain the genetic diversity. All the species showed one common band but variability was restricted to minor bands. Based on the number of bands and Rf values, these species were categorised into six groups. Fifty mulberry landraces and popular varieties were subjected to Isozyme analysis using three enzyme system namely anodal peroxidase, acid phosphotase and cathodal peroxidase to study the extent of genetic diversity.

Peroxidase banding pattern showed greater variability. *M. laevigata* is distributed in different parts of the country both in wild and managed habitats. In order to assess the genetic diversity in 32 *M. laevigata* explored collection were subjected to isozyme analyses. Peroxidase, ß- esterase and Cathodal peroxidase enzyme systems showed wider genetic polymorphism. Acid phosphotase enzyme again did not show any genetic polymorphism among 32 *M.laevigata* accessions.

Isozyme banding pattern did not show any correlation with geographical distribution of collections. However, *M.laevigata* collections from north-eastern India revealed maximum genetic diversity at isozymic loci indicating potentiality for further exploration and collection. Polymorphism at Isozyme loci is comparatively low in mulberry as is the case in most of the cultivated crops. Hence, there is a need to take up more number of enzyme systems for Isozyme analyses to obtain conclusive result.

DNA finger printing

Molecular characterisation of mulberry germplasm was initiated in collaboration with Seri Biotech Research Laboratory (SBRL), Bangalore. Initially, protocol for mulberry DNA extraction, RAPD and Inter SSR- PCR analysis was standardised. Fifteen morphological distinct species were analysed using RAPD and Inter SSR-PCR assays. In RAPD assay, 19 oligonucleotide primers were utilised. These analyses also identified few species-specific amplification products. Inter SSR-PCR analyses were carried out in 15 mulberry species using four primers. However, the species-specific markers have to be studied over populations of individual species so as identify the unknown species.

The RAPD analyses of 46 mulberry varieties using five oligonucleotide primers revealed polymorphism suggesting the genetic variation. However, clustering pattern did not show correlation with regard to regional/ geographical distribution of these varieties. National Botanical Research Institute, Luknow studied nine mulberry germplasm varieties (triploids, species and popular varieties) using 23 arbitrary sequence decamer primers for RAPD and three minisatellite core sequence primers for DAMD polymorphism. Nearly 85% RAPD and 91% DAMD bands were polymorphic across nine varieties differentiating ployploids and popular varieties.

Genus Morus L.

Ever since sex in plants first demonstrated in mulberry, the sexual polymorphism in genus Morus continues to be enigmatic and forms the study material for many classical taxonomists. Linnaeus for the first time established genus Morus with 7 species. Hooker placed genus Morus L. under Urticales and tribe Moreae. Takhatajan rightly placed genus Morus under Moraceae family of order Urticales which is considered to be advanced in the woody flowering plants. Different authors have classified genus Morus based on morphological and phenological characters.

The taxonomical status of genus Morus L. has been critically reviewed by Sanjappa, Tewary *et al.*, Ramesh and Basavaiah. So far more than 150 species (vide Index Kewensis) have been described under genus Morus, however, most of them have been now reduced as synonymous or varieties of the same species. Presently there are about 68 recognised species endemic to 13 nations. The comprehensive classification of genus Morus with 25 species by Koidzumi mainly based on style length and stigma hairiness though holds good but there is no place for male unisexual accessions in his classification.

Tikader *et al.*, and Tikader and Rao, used morpho-reproductive characters of quantitative nature for divergence analysis of ten Morus species and found that *Morus alba* formed separate cluster for all the traits studied and reproductive characters have contributed maximum forming seven clusters. Dandin *et al.* studied the chromosome configuration and phylogenetic relations of different varieties of four mulberry species. Recently, using molecular techniques with four ISSR and 19 octamer random primers, the species-specific markers in 15 mulberry species were identified.

However, these markers are to be linked and correlated to the qualitative and quantitative traits. Reviewing the taxonomical status of genus Morus based on literature available, it is clear that various authors have classified genus Morus with limited number of mulberry species of particular geographical origin. Nevertheless, an agreeable system with diagnostic keys of different species for correct classification of genus Morus L. is yet to emerge out. There is an urgent need to procure the wild relatives of Morus species from different geographical endemic regions through survey and explorations and study the numerical taxonomy using various descriptors including molecular techniques for the revision of genus Morus on world basis.

Evaluation of Mulberry Germplasm

The assessment of genetic potentiality of mulberry germplasm is a vital for crop improvement programmes. The breeders mostly depend on the evaluated material for choosing parental lines. Evaluation of germplasm is mainly done for propagation, growth and yield, nutritional quality, biotic and abiotic stress tolerance and region and season specific performance.

Mulberry is chiefly propagated through vegetative stem cuttings to maintain the genetic potentiality of evolved varieties. Rooting behaviour of mulberry accessions is an important trait in the breeding programme. In the temperate belt, usually the introduced exotic varieties such as Goshoerami and Kosen propagated through grafts, on the stocks raised as seedlings of local varieties. Hence, scion and stock compatibility of different mulberry accessions are to be studied for utilisation of temperate varieties. Earlier attempts were made at CSR&TI, Mysore with 61 accessions taking root length and root-shoot ratio.

Screening of rooting ability in cuttings of different ploidy levels of mulberry accessions have indicated that triploids are better rooters followed by diploids and then tetraploids. Correlation of rooting traits with growth parameters was worked out in 31 accessions. Morus multicaulis is a good rooter with high yield potential for exploitation. Sujathamma and Dandin, studied sprouting and survival of 25 germplasm varieties under arid zone. At CSGRC, Hosur, propagation studies related to survival and rooting behaviour were conducted in 530 accessions (indigenous-390 and exotic-140) during spring and rainy seasons of 1996-99 following RBD design. Exotic varieties mainly from temperate belt have showed poor rooting ability.

Field Gene Bank Evaluation

The yield attributing parameters of mulberry accessions are first to be assessed in the field gene bank to group and shortlist mulberry accessions for further screening through primary and final yield trials. In mulberry, there is a paucity of information available on the evaluation of germplasm as dwarf tree for different growth parameters. At CSGRC, Hosur during the year 1993-1997, 628 mulberry accessions (Indigenous-455 and Exotic-173) were assessed for their yield attributing characters for 16 growth and yield parameters. The analyses of variance indicated that both the seasons and accessions are significant.

Exotic varieties exhibited higher mean values for different leaf characters whereas indigenous accessions showed higher mean values for shoot characters. Evaluation of mulberry germplasm suited to farmers package of practices is an added advantage for selection of superior genotypes. Hence, an experiment was conducted at CSGRC, Hosur with 316 accessions (234 indigenous + 78 exotic + 4 checks) for preliminary evaluation for growth and yield parameters under ARBD. This would facilitate to shortlist promising genotypes for further screening at breeders level after evaluation for quality parameters through bioassay studies. A total of 45 mulberry accessions were finally selected for further evaluation. Bioassay (full rearing test) of these 45 accessions has been conducted to select promising accessions.

The contribution of mulberry leaf for successful harvest of cocoon crop is well recognised. Varietal differences in moisture, protein, mineral and sugar content in mulberry germplasm are reported by various authors. At CSGRC, Hosur, biochemical characterisation was completed in 628 accessions using the leaf samples collected from base collection. Chlorophyll-a, Chlorophyll-b, carbohydrate and protein (water soluble) content were estimated as per standard procedures. These studies have indicated the potentiality of exotic varieties in relation to quality parameters.

Evaluation for Biotic Stress

Mulberry is affected by large number of diseases caused by fungi, bacteria, viruses and nematodes and pests. The diseases besides causing heavy crop loss (10 to 30%) also reduce the leaf quality. One of the most promising and economic ways of evaluating the disease loss in mulberry is to evaluate germplasm collections in the hot spots, which would also help in identifying genotypes for breeding disease resistant varieties. Ninety-two mulberry accessions were screened for various diseases at CSR&TI, Mysore of which 17 were found to be resistant to leaf spot and mildew diseases with 5% disease incidence.

Only three accessions were moderately resistant to rust disease. The varieties Kaliakuthai and Bilidevalaya were found to be completely resistant to leaf spot/ powdery mildew. At CSGRC, Hosur preliminary screening of 328 mulberry accessions against three major diseases such as powdery mildew, leaf rust and leaf spot was carried out in the field gene bank to select disease tolerant varieties. However, these accessions are to be tested in the hot spots for their conformation.

Evaluation for Abiotic Stress

It is estimated that various abiotic stress conditions such as moisture (drought), salinity and alkalinity result in 50-60% yield loss in case of mulberry, thereby affecting the economics of sericulture.

Drought Resistant Mulberry Cultivars

Some researchers have successfully demonstrated the need of wider genetic base and importance of mulberry germplasm characterisation and preliminary evaluation with reference to leaf anatomical characters (high thickness of leaf, smaller stomata in lower frequency, thicker cuticle, and higher spongy palisade layer ratio), rooting parameters (length of root, dry weight of root, root shoot ratio by length and weight) and yield characteristics (number and length of branches, leaf weight and area, internodal distance, leaf shoot ratio, moisture content and retention capacity of leaves) for identification and evolution of drought resistant mulberry cultivars. These studies have resulted in releasing several drought tolerant varieties such as S-13 etc.

Field studies have indicated that chlorophyll stability index is an important drought resistance parameter since the increase in level of total chlorophyll (a & b) is found to be minimal under severe stress condition in potential drought resistant varieties of mulberry. Studies on drought induced cellular events in mulberry have indicated the increased activity of proline and glutathione. Such methods of quantification of bio-chemicals can be extended to preliminary screening of germplasm accessions for drought resistance.

Earlier observations in Mulberry though have indicated marginal to significant level increase of osmotins such as proline and glycine betaine, however without significant difference between drought susceptible and drought resistant genotypes. Hence, suggesting that level of such osmotins do not lead to any meaningful conclusion to be associated with stress resistance. Mulberry germplasm accessions selected through different morpho-physiological, propagation and yield evaluation characters at CSGRC can be subjected to preliminary pot screening test under stress and non-stress conditions to assess drought tolerance based on wilting percent and wilting recovery and other growth characteristics of saplings. The tolerant genotypes are to be tested in the hot spots for selection of drought tolerant varieties.

Alkaline or Acidic Tolerant Mulberries

Mulberry is cultivated often in soils, which are alkaline or acidic as in some parts of Karnataka (Mysore district) and Tamil Nadu (Salem district) of Southern plains and West Bengal and Bihar in Eastern plains. Mulberry can tolerate pH range of 5 to 8, however it grows best in the soils having pH around 6.5. Empirical testing in hot spots for saline or acidic soil is the best way to screen genotypes tolerant to such soils.

CSGRC, Hosur in collaboration with its regional net working stations of CSB units has to take up the evaluation trials of the promising mulberry germplasm accessions to select alkaline tolerant varieties. The hot spots identified in Southern plains for alkaline soils are Kinakanahalli under RSRS, Chamarajanagar (pH 8.5-9.5) and RSRS, Salem in Tamil Nadu. Similarly for acidic soils, Kishanganj (Bihar) is a hot spot with pH around 5.25. CSRTI, Mysore has conducted yield trials at Kinakanahalli under RSRS, Chamarajanagar resulted in identification of alkaline resistant varieties.

Salinity

Salinity is one of the limiting factors of crop yield in nearly one third of irrigated land. Sericulture productivity is more dependent on irrigation or partial irrigation since mulberry responds well to irrigation; as evidenced by intensive cultivation in Southern and Eastern plains of India, where nearly 40% of mulberry cultivation is under irrigation. Salinity influence growth in mulberry through reduction in root growth at lowers levels and at higher level it has greater effect through reduction of shoot growth. Above 30 mM of salinity drastic reduction in growth of both shoot and root are reported.

Salinity also strongly decreases the protein, starch, and sucrose levels in leaves. Protocol for screening of large number of germplasm accession of saline tolerant lines in mulberry, through *in vitro* screening, using lower concentrations of NaCl at 0.1% and 0.2% level to know the effect on rooting has been suggested. Rooting parameters such as number of roots/plant, average root length and length of longest root are considered important in such screening.

Raising of saplings from mulberry cuttings in hydroponics culture solutions with different concentrations of NaCl have been suggested as a method for screening large number of mulberry accessions for saline tolerance. The specific effects of salinity on rooting ability, early growth and level of chemical constituents are assessed in this procedure. Level of increase in total chlorophyll content in particular and total proline, glycine, betaine and peroxidase in saline conditions are known to be useful markers.

Utilisation and Development of Mulberry Resources

"Sustainable Use", means, the use of components of biological diversity in such a manner and at a such a rate that does not lead to long-term decline of the biological diversity thereby maintaining its potential to meet the needs and aspirations of present and future. The most effective way to conserve mulberry germplasm resources is to put them in continuous use meeting competing demands of the society. Hence, multipurpose uses of mulberry germplasm have to be greatly exploited for their sustainable conservation.

Mulberry improvement programmes utilising genetic resources in India has travelled a long way. Mulberry genetic resources have been used in the breeding programmes mainly through controlled hybridisation, polyploidy breeding, mutation and selection. The Indian sericulture in early 1950s-1960s has witnessed the use of local landraces belong to species Morus indica like Kajli in West Bengal and Mysore local in Karnataka with highly dissected leaves. During 1960s introduction of broad entire leaved varieties like Gosheorami, Tsukasaguawa and Ichinose from Japan in the temperate region of Jammu and Kashmir and Matigara black in north eastern India under wet tropical climate, which have marginal, increase in the leaf yield.

The mulberry improvement programmes utilising germplasm resources were started in India during 1970s after selection of open pollinated hybrid Kanva-2 from Mysore local and the clonal selection Ber-S1 from Mandalaya (Introduction from Burma). Utilisation of local landrace Berhampore local in West Bengal probably paved the way for developing high yielding varieties like S-30, S-36, S-41 and S-54 developed through mutation breeding techniques and V-1 through controlled hybridisation between S-30 (Parent Berhampore local) × C-776 in South India. This resulted in recognition of potentiality of diverse germplasm resources with wider genetic base for crop improvement programmes.

In West Bengal, Mandalaya (Morus alba) introduction from Burma has been extensively used for polyploidy breeding using local Morus indica (HP) as a female parent because of its copious female flowering and a series of triploids and tetraploids have been evolved. The most promising triploids from West Bengal like Tr-4, Tr-8 and Tr-10 released for hilly regions and C-1730 and S-1635 for irrigated regions. The success of triploid varieties for commercial cultivation in west Bengal has attracted the breeders and prompted them to use high yielding diploid parents like S799, Chinese white, CRS for developing triploids.

The promising triploids like S-1635, C-1730 are released in West Bengal during 1990s. The conceptual utilisation of different mulberry (Morus species) in crop improvement resulted many high yielding varieties in many countries. Out of 68 recognised species of genus Morus, approximately 35 species so far identified, which possess genes for exploitation for quantity and quality improvement of leaf yield in mulberry. The potentiality of species *Morus latifolia/Morus multicaulis* in crop improvement programmes have been well recognised by many breeders in China and Japan.

In India *Morus multicaulis, M.latifolia* (var. Kosen) has been used along with *Morus indica* (var. Black cherry) for developing high yielding varieties like C-763, C-776 and BC-2-59 through controlled hybridisation and back cross breeding in West Bengal during 1980s. Utilisation of diverse germplasm resources for abiotic stress tolerance for wide range of environments has yielded varieties like S-13, RFS-135 and RFS-175 for semi-irrigated conditions.

Realising the increased use of mulberry germplasm resources for crop improvement programmes, CSGRC, Hosur being the nodal Institute for mulberry germplasm conservation and utilisation in India taken up massive experiments like characterisation and evaluation of the entire germplasm holdings at national field gene bank to group and short listing of trait specific promising accessions which will facilitate breeder/users.

Mulberry is fast growing tree and produce large quantity of renewable biomass in the form of branches, shoots, leaves and fruits. One hectare of mulberry garden yields up to 60-70 tonnes/ha/year of green leaf and 12.1 tonnes of mulberry sticks (V1 mulberry variety). The energy generated per hectare is 27 940k calories @ 4 600 calories/kg of mulberry wood. It promotes employment generation of 12 persons/ha/year. The green leaf with rich protein content (up to 28% on dry wt. basis) is used as fodder for cattle in many parts of India.

The wood is used for making furniture, carving, toys making, sports goods, agriculture implements and papermaking. The fruits are eaten fresh and also used for jam, jelly and wine preparations. In the under-utilised fruit genetic resources category mulberry has a great potential for exploitation. It can be cultivated in diverse habitat both in temperate and tropical climate. Mulberry fruits are having 100% edible portion with 80% of moisture and rich in protein, iron, phosphorus, calcium, minerals and vitamins like carotenes, thiamine, riboflavin, niacin and vitamin C. An average single mulberry plant under dwarf tree condition can yield about 6 to10 quintals/acre/year of fresh mulberry fruit (estimated yield) and this may vary with species and cultural practices.

Medicinal Purposes

It is estimated by World Health Organisation (WHO) that 80% of world population still relies on traditional system for primary health care. The various parts of mulberry plant are used in ayurvedic preparations The root bark, twigs and fruits which contain phenolic compounds, morasin and cyclomorasin are used as restorative, tonic, pectoral, diuretic and are prescribed to treat cough, asthma, phthisis and other chest complaints, dropsy and rheumatism. The decoction of the leaves possess blood purifying properties as a febrifuge, diuretic and galactugogue and stringent.

The syrup made from fresh fruits is used as laxative, refrigerant in fevers and as an expectorant in cough and sore throats dyspepsia and Melancholia. The root extracts is having hypoglycaemic properties and used in treatment of diabetes. The root bark is used as a purgative and vermifuge. Mulberry root juice is administered to patients with high blood pressure. The shoot contains latex, which is used as plaster for sores and for preparation of skin ointments.

Trade and Global Exchange of Germplasm

The relation between trade and global exchange of germplasm became more intricate under Intellectual Property Rights (IPR) regime by TRIPS agreement under GATT. The Indian government preferred to use Sui generis option because of three major advantages i.e. flexibility, better protection of farmer rights and stronger researchers exemption. Registration of germplasm is an important aspect for protection of breeders right under national and international agricultural/sericultural research systems.

Outstanding varieties of mulberry are released for commercial exploitation after authorisation. However, many promising germplasm do not get authorised due to one or other limitation in spite of their outstanding performance. Registration of such germplasm is required so as to give due recognition to the scientists who have developed the material. Lack of formal recognition of such useful germplasm materials discourages the scientist/breeders for supply of these materials to others for utilisation. Consequently, such valuable materials remain under utilised or may even get lost.

Moreover, there is urgent need to document such materials in wake of GATT and IPR regime at national and international level. Keeping these points in view, CSGRC, Hosur has developed a set of guidelines for registration of sericultural germplasm under National Sericultural Research System of CSB.

In India there is a single window system for germplasm exchange of small samples of PGR for research purposes and it is negotiated through NBPGR, New Delhi which is the nodal agency to plan, co-ordinate, execute and monitor PGR management in India under Indian Council of Agriculture Research (ICAR) in the Department of Agriculture Research and Development (DARE). In India, the Directorate of Plant Protection, Quarantine and Storage (DPPQS), Ministry of Agriculture, New Delhi monitor plant quarantine regulations with its regional network centers at airports, seaports and land frontiers.

Under enactment of new policy on seed development, any plant breeder intending to import planting material of mulberry including in vitro cultures has to get import permit addressing to Director NBPGR, New Delhi on prescribed format. The Phytosanitory certificate which is to be met by the supplier along with the sample ensuring that the material is not coated treated with chemical pesticide and the envelop to be addressed to the Director, NBPGR New Delhi who takes the delivery in full of the consignment and contain required for quarantine check.

The material so introduced shall, after quarantine clearance and National registration (accessioning), will be made available to the indenter. Some of the introduced material may also be examined through post entry quarantine growing under controlled conditions for hidden pest/pathogen. The export of germplasm to any country should also be regulated through NBPGR, New Delhi. The material will be completely processed for quarantine check and issued a Phytosanitory certificate.

Quarantine process includes examination of plant material for the presence of any pest/pathogen, unwanted weeds, through pest detection techniques like visual microscopic examination, X ray radiography, soaking seeds, sieving, staining, blotter method, growing out test, infectivity test and Enzyme Linked Immuno-sorbent Assay (ELISA). Under Article 19.3 of CBD, the potential risk posed by transboundry movement of Living Modified Organs (LMOs), which may have an adverse effect on biodiversity and its components will also be taken into account if necessary.

The International Treaty of Plant Genetic Resources established " Multilateral System" for facilitated access to a specified list of PGR for food and Agriculture, benefit sharing in the areas of exchange of information, technology transfer and capacity building. At present 60 plant genera, which includes approximately 35 crops complexes, are listed in the Multilateral System. Paradoxically, mulberry is not included in the Multilateral System.

Mulberry Germplasm Database

Conservation of mulberry germplasm on long-term basis requires huge resources and manpower and must be accompanied by effective utilisation of germplasm resources on sustainable basis to justify long-term investments. In the backdrop of dwindling natural seribiodiversity with changing global climatic scenario, it is assumed that all genetic resources are potentially valuable and to be conserved for posterity. Present value of the existing resources cannot be assessed because of future developing technologies. This ends up in accumulating vast genetic resources in the field gene bank year by year. Hence, it is imperative to analyse the cost of the biodiversity conservation and its sustainable utilisation.

The documentation acts as sources of information to assist in the setting of the priorities, planning and operation of germplasm management. For effective information management of PGR, a complete inventory of the mulberry genetic resources of the nation with their Passport has to be developed. The information can be useful, meaningful and economical when it is properly analysed and logically organised to make it user friendly, easily available and retrievable.

Realising the importance of information system in the germplasm management, CSGRC, Hosur developed an information system named Mulberry Germplasm Information System (MGIS) in international accepted language English incorporating various facilities. This system will work on MS-DOS platform. In order to MGIS more meaningful and user friendly, the standard data format in terms of terminology and measurements with integrated database techniques were used. User-friendly menu driven facilities have been incorporated for easy data viewing, information retrieval, updating and report generation. Each file is linked to the National accessioning number which is a unique number assigned to each accession serving as primary key.

The query module has the facility for identification and selection of accessions on the basis of certain desirable criteria up to 15 quantitative/qualitative characters has been developed as a separate module to facilitate breeders/users for utilisation of information. The system is flexible enough to allow modification and insertion of new modules as per the changing needs of the scientists. In future, MGIS will be developed in Window platform using GUI (Graphic User Interface) technology.

Methodology for Morphological Characterisation

The morphological data has to be recorded in the base collection of three years old plants, maintained as high bush/dwarf tree with spacing of 2.4 m ´ 2.4 m distance and crown height of 1.5 m with recommended cultural practices for high bush maintenance. The plants have to be pruned twice in a year i.e. in Feb-Mar and July-Aug for recording of data. The different morphological data has to be recorded after 75-90 day of pruning in both the seasons.

Methodology for Reproductive Characterisation

The reproductive characterisation has to be recorded in the main flowering season (February-March). Fully bloomed catkins of different accessions should be randomly collected and the length and breadth and the pedicel length of the different catkins should be measured. Number of flowers/catkin of different sexes should be counted by dissecting the individual flowers around the rachis of the inflorescence. Style length and stigma length, nature and type of the female flowers are to be recorded by dissecting them under stereomicroscope at receptive stage.

The stigma is receptive from 8-12 days after floral initiation. The male catkins should be collected at 10 am in the morning at the time of anthesis in plastic bags and the stamen and the anther length should be measured under the stereomicroscope using ocular eyepiece micrometer. For studying the pollen diameter, viability and number of pollen germ pores, the mature pollen grains should be dusted on a clean slide from the male flowers and stained with 2% acetocarmine solution for 30 minuets and studied under the microscope. Fully ripened sorosis should be collected and different fruit characters like fruit length, breadth, pedicel length, weight, colour and taste to be recorded.

For studying the different seed characters the fruits are to be soaked in water for over night and the seeds are properly cleaned with water. The floated seeds are rejected and only the healthy seeds, which are deposited in the bottom, are separated and dried on blotting paper. For calculating the seed set percentage prior to washing the number of acnes per fruit and number of seeds per fruit is to be recorded separately.

Methodology for Leaf Histology

Fully expanded leaves from 7th to 9th position in descending order from three months old pruned shoots in base collection are to be considered for the study. Small rectangular pieces should be taken from the central portion of the leaf blade avoiding veins and veinlets and preserve in the FAA solution. Quick fix replicas from adaxial and abaxial surfaces of the preserved leaf samples to be used for studying the stomatal size and idioblast frequency. While measuring the stomatal sizes care to be taken to avoid giant sunken stomata.

For counting the number of chloroplast in the stomata, fresh epidermal peelings from 5th to 7th position leaf from the top of the branch in a descending order to be considered. For measuring the idioblast projection length, cystolith length, breadth and thickness of different layers such as cuticular, epidermal, palisade, spongy and total leaf thickness, thin hand sections of the preserved leaf material has to be taken and stained with 1% safranin and mounted in 50% glycerine. For all the characters three leaves per plant and three plants per accession are to be considered.

Methodology for Cytological Characters

The detailed karyotype studies can be made by selecting the well spread mitotic plates and all the chromosome measurements like short arm length, long arm length, total chromatin length, satellite bearing chromosomes are to be made from the camera Lucida drawings or from photomicrographs in mm and then converted to microns. All the measurements are to be tabulated separately and homologous pairs are to be arranged in the descending order of total length.

Methodology for Growth and Yield Attributing Parameters

The plants are to be maintained in row with 2.4 × 2.4 m spacing, as dwarf tree with crown height of 1.5 m. Recommended cultural operations for dwarf tree maintenance has to be followed. The plants are to be pruned twice in year in spring (February-Mar) and autumn (June-July). The data on different growth parameters has to be recorded when the plants attained three years of age and after 75-90 days of pruning considering at least three plants/accession. Moisture content of the excised leaves and moisture retention capacity (after six hours storage) has to be determined using equal number of top, medium and mature leaves by gravimetric method by using the formula.

Global Sharing of Germplasm Information

Mulberry plant genetic diversity is unevenly distributed

over the globe and most of its diversity is concentrated in the tropical and subtropical geographical zones. No single country is self sufficient in PGR. The development of the global system on PGR began after the establishment of Communication on Genetic Resources for Food and Agriculture (CGRFA) by FAO.

The main objectives are promotion and utilisation of PGR by providing flexible framework for sharing information and developing networking activities, interaction, agreements, code of conduct, scientific standards and instruments for PGR management. The International Undertaking on Plant Genetic Resources (IUPGR) and CBD advocates the utilisation of PGR with benefit sharing, on the contrary TRIPS and UPOV (Union for Protection of New Varieties of Plants) stressed on demand on ownership rights, which restricts the global exchange of PGR. Morus diversity is rich in South Asian countries mainly in Japan, China, India, Thailand, Indonesia and Korea.

The native genes of Morus species of continental America are untapped for sericulture development. The sharing of mulberry germplasm information and vis-à-vis utilisation of genetic resources at global level is hitherto running at slow pace due to non coherence of international understandings over the issues related to plant genetic diversity and utilisation. Today mulberry germplasm resources are available in 30 countries or more. There is an urgent need to establish the International Networking on Mulberry Germplasm Management and Utilisation at global level. Japan is assisting developing countries through its JICA programme. However, special emphasis to be given for international training programmes on mulberry germplasm management under JICA.

8

Potential of Mulberry Foliage

A major constraint to animal production in developing countries is the scarcity and fluctuating quantity and quality of the year-round feed supply. These countries experience serious shortages in animal feeds of the conventional type. Grains are required almost exclusively for human consumption. By the year 2020, world population is expected to reach eight billion and most of the population growth will occur in developing countries. With increasing demand for livestock products as a result of rapid growth in the world economies and shrinking land area, future hopes of feeding the millions and safeguarding their food security will depend on better utilisation of unconventional feed resources that do not compete with food for human beings.

In addition, over-exploitation of soil resources has resulted in wide scale land degradation. This calls not only for better utilisation of already known unconventional feed resources but also for the identification and introduction of new and lesser known plants capable of growing in marginal soils. The propagation of such plants can play a vital role in the control of soil erosion, bring economic benefits to farmers, create jobs and bridge the wide gap between supply and demand for animal feeds. These factors would also lead to diversification in traditional agriculture and conservation of biodiversity through sustainable utilisation of natural resources.

The rearing of livestock is an integral part of the economy of most of the developing countries. The economy of the Indian farmer is based to a large extent on animal production. However, production is quite low as the animal is undernourished for a significant part of the year. The low productivity is exacerbated by long calving intervals and a late age of puberty.

Crop residues and low quality forages are the major feed resources. The primary constraint to ruminant production on such feeds is the low efficiency of feed utilisation. Livestock development over the past three decades has been mainly directed towards satisfying the rapidly increasing demand for milk and meat in the urban centres. The resource-poor small farmers produce a major part of this milk and meat. Appropriate technologies to improve the performance of locally available animal and feed resources within the rural system are lacking.

Animal productivity can be increased by the introduction of low cost technologies that improve the current systems of husbandry. Acceptable and successful feeding systems are those that are simple, practical, consistently reproducible and within the limits of the farmer's capacity and available resources. Fodder, the mainstay for livestock rearing, is cultivated on only four percent of the total cultivable land in India and this figure has remained more or less static for the last three decades. Because of the increase in human population, demand for food grains and other cash crops is increasing, resulting in a smaller area for fodder cultivation. The denuded grasslands, pastures, forest openings and the forests are the major source of herbage for livestock.

The 1993 draft report of the Policy Advisory Group on Integrated Grazing Policy, Ministry of Environment and Forests, Government of India indicated the deficit was 584 and 745 million tonnes of dry and green fodder respectively. The present level of availability of animal feeds indicates a deficiency of about 19 percent DM, 55 percent DCP and 28 percent TDN for feeding the existing livestock population. The increasing human population, limited land holdings and the growing requirements of food grains for the human population have a direct impact on livestock production.

Recent nutritional research has demonstrated the possibility of a large increase in animal production that can be achieved by alterations to the feed base. Production can be increased by up to fivefold by providing the critical catalytic nutrients that are deficient in the diet and by balancing the availability of nutrients closer to requirements. In the rangelands, particularly in the semi-arid areas, tree forages, seeds and pods represent by far the greatest potential source of protein meals.

In more recent times, trees and shrubs have been introduced into cropping and grazing systems to provide green fodder that is high in protein to supplement the available low protein forage. These are grown in banks or hedges, between crops or as a component of pastures and as shade trees. Multipurpose trees can make a significant contribution to agricultural systems by improving soil fertility and providing a variety of useful products, including valuable forage and wood.

The feeding value of low-quality roughages and grasses can be greatly improved by the foliage from trees, which can be integrated directly with pastures and in fences. In some cases, pure stands of forage shrubs and trees can be the best option to intensify animal production replacing traditional low performing grass-based systems. The potential roles of tree foliage in ruminant nutrition are: a) a high quality and high digestibility biomass resource, available in and around the farm; b) a supplement to provide nutrients deficient in the diet resulting in enhancement of microbial growth and digestion of cellulosic biomass in the rumen; c) a source of undegradable protein; d) a source of vitamins and minerals to complement deficiencies in the basal fed resource, and e) a reduction in the requirements for purchased concentrates and as a result the decrease in a cost of feeding.

Tree forages form an integral part of ruminant feeds in the high altitudes of the Himachal Pradesh, Jammu and Kashmir and Uttar Pradesh states of India. The use of tree forages as components of diets is a widespread practice in many tropical countries. Considerable diversity exists in the type of forage supplements of value, particularly to ruminants. Recognition of the potential of tree foliage to produce considerable amounts of high protein biomass has led to the development of animal farming systems that integrate the use of tree foliages with local bulky feed resources. In order to determine the suitability of trees/shrubs as components of ruminant fibrous diets, knowledge is required in many areas, including:

- the capacity and ability of the tree to regenerate foliage when grazed or harvested;
- the feeding behaviour of animals when confronted with tree forages;

- the voluntary intake of tree foliage under different environmental conditions;
- the adaptation of trees to the local conditions and their potential to become weeds;
- the growth pattern of trees/shrubs in relation to crops or pasture;
- the required soil pH characteristics and nutrient status;
- the nutritive value of the foliage and its change with harvesting, grazing or cultivation.

The mulberry plant in the tropical belt is grown as a low bush while it is grown as a high bush in temperate regions. In tropical conditions, individual leaf and branch harvest yields 10 to 30 tonnes/ha/year, while shoot harvesting in temperate regions has a leaf yield of 25 to 30 tonnes/ha/year. The percentage of moisture, protein and carbohydrates is higher in temperate regions when compared to the tropics. Mulberry is a monoecious, occasionally dioecious shrub or moderate-sized tree with a fairly cylindrical straight bole, up to 3.0 m high and 1.8 m in girth. Leaves are very variable, ovate or broadly ovate, serrate or crenate-serrate, and often deeply lobed. The plant is frost hardy but liable to wind damage. It regenerates itself naturally from seeds that are dispersed by birds and to a limited extent by jackals and human beings. It can be propagated artificially by seeds or cuttings.

It grows rapidly in the early stages and reaches maturity at an early age; the growth rate falls off rapidly after approximately ten years. Mulberry coppices vigorously and pollards well. When grown close in plantations, the tree develops a long clean bole. At present in many other parts of the world, mulberry leaves are predominantly used for silk production, and therefore it will be pertinent to mention the status of silk production from mulberry in India. This might encourage workers to develop strategies to integrate silk and livestock production at the village level, resulting in higher incomes for farmers.

Silk Production

Numerous types of mulberry are under cultivation in various silk-producing countries of the world; the types differ in their adaptability to various soils and climates, resistance to diseases, food value of the leaf crop for the silkworm and suitability for use as stock or scion in grafting. In Japan, the world's major silk-producing country, approximately 700 types of mulberry are known to exist, of which 21 have been selected for extensive cultivation. Some of these types are adaptable to a wide range of climatic conditions.

The most important type of mulberry grown in India for rearing silkworms is *Morus alba* var. *multicaulis* Loud, which is a native of China or the Philippines. It is fast-growing and adapted for cultivation as a field crop producing large, tender and thick leaves. *M alba* var. *atropurpurea* (Shahtut), also a native of China, is cultivated widely as a tree crop for its large, cylindrical, dark purple, succulent fruits. It is also fast-growing and yields large thick leaves, which are at the same time smooth, tender and succulent fruit. *Atropurpurea* is recommended for cultivation along the borders of bush plantations in Bengal.

Mulberry is grown on an extensive scale in various parts of India, particularly in Mysore, West Bengal and Jammu and Kashmir for its leaf, which constitutes food for the silkworm. Its cultivation is an integral part of the sericultural industry. In some areas, it has run wild and spread. It is also grown as a roadside and avenue tree. In the Himalayas, it reaches a height of about 1 200 m. In the hills, it is mostly confined to stream beds or places where sufficient moisture is available. It does not grow on dry slopes or shallow soils where moisture is a limiting factor. The old leaves are shed in November to December and the trees are leafless during the winter season. The new leaves appear in March to April depending upon the climate of the locality. Mulberry tolerates shade and it can with advantage be grown as an understory with other light-demanding species.

In Mysore and West Bengal, mulberry is grown as a field crop and the leaves are harvested several times a year to feed the multivoltine races of silkworm. In Jammu and Kashmir, it is grown as a tree and the leaves are lopped only in one season for rearing the univoltine races of the worm. A system of growing dwarf grafted trees or "high bushes" has recently been tried in West Bengal. Mulberry is also grown to a small extent in the Punjab, Himachal Pradesh, Uttar Pradesh, Madhya Pradesh, Bihar, Orissa, Assam, Manipur and Andhra Pradesh states of India, where small quantities of silk are produced. The largest area is in Mysore, which accounts for more than 75 percent of the total mulberry raw silk production in the country.

Young leaves that have attained full size are best suited for feeding silkworm larvae. The composition of the leaves varies according to the type, degree of maturity and the soil in which the plants are grown. The content of protein and soluble sugars in leaves decreases with the maturity of leaves whereas fibre, fat and ash constituents increase. Young leaves are more acidic than older ones. An analysis of leaves collected from different silk-producing localities in India gave the following ranges of values (percentage in DM): CP 16-39, soluble sugars 7.6-26 and ash 8-17.

India has the advantage of producing of all the four known commercial silk varieties in the world, i.e. *mulberry, tusser, eri* and *muga.* Mulberry silk is domesticated while the rest are wild. India produces currently about 14 050 tonnes of silk of which mulberry alone accounts for more than 90 percent of the total silk production, practised in over 60 000 villages in the country. Mulberry sericulture is well suited for marginal, small and landless farmers since it has several advantages over other crops in the rural sector. It is estimated that one hectare of irrigated mulberry will generate employment for about 13 people starting from mulberry cultivation to trading throughout the year and that, in value addition, 48.3 percent goes to the farmer; 21.6 percent to traders and the rest to weavers, reelers, twisters and dyers.

In Himachal Pradesh, mulberry is grown as a shade tree on wastelands, roadsides, borders of fields and around farmers' houses. Himachal Pradesh is predominantly a univoltine/bivoltine silk-producing state. Mulberry leaves for silkworm rearing are collected from trees grown on the border of the fields, around the houses and on the wastelands under rainfed conditions. Family labour is employed in sericulture; women play a major role. Sericulture is adopted mainly by small and marginal farmers and landless labourers as a subsidiary occupation.

The preferential food value of mulberry leaf for silkworm larvae is attributed to the presence of three stimulant factors in it, i.e. an attractant, a biting factor and a swallowing factor. The substances that attract the larvae to the leaves have been identified as citral, linalyl acetate, linalol, terpinyl acetate and hexenol. Sitosterol, together with some sterols and a water-soluble substance, is the main factor that stimulates the biting action.

Animal Feeding

The leaf fodder of mulberry is reported to be of good quality and can be profitably utilised as a supplement to poor quality roughages. Leaf yield varies according to the fertility of the soil, irrigation and frequency of plucking of the leaves. In West Bengal, one hectare of well manured and irrigated plantation can yield about 19 to 28 tonnes of leaves in five pluckings. On a DM basis, the leaves contained 15.0-27.6 percent CP, 2.3-8.0 percent ether extract (EE), 9.1-15.3 percent crude fibre (CF), 48.0-49.7 percent nitrogen free extract (NFE), 63.3 percent total carbohydrates, 14.3-22.9 percent ash, 2.42-4.71 percent Ca, 0.23-0.97 percent P, 0.196 percent S, 1.66-3.25 percent K, 350-840 ppm Fe. The cell wall constituents were: NDF 33-46 percent, ADF 28-35 percent, hemicellulose 5-10 percent, cellulose 19-25 percent, and lignin approximately 11 percent. The content of total phenols was very low, and tannins by the protein precipitation capacity method were not detectable.

A prolamine has been separated from the alcoholic extracts of mulberry leaves and it forms the principal protein of the leaves. The nitrogen (N) distribution in a preparation containing 12.6 percent N was as follows: HCl-insoluble N 0.50; humin N 0.45; amide N 0.96; diamino acid N 1.74; and monoamino acid N 7.89 percent. Protein preparations from young mulberry leaves form an excellent supplement to protein-deficient diets. Non-protein nitrogen accounts for approximately 22 percent of the total N in young leaves and approximately 14 percent in mature leaves. The amino acids identified in the free form are: phenylalanine, leucine, valine, tyearosine, proline, alanine, glutamic acid, glycine, serine, arginine, aspartic acid, cystine, threonine, pipecolic acid and 5-hydroxy pipecolic acid.

The mulberry leaves are thus rich in CP, EE, calcium and ascorbic acid and low in CF. They also contain carotene, vitamin B1, folic acid, folinic acid and vitamin D. The presence of glutathione in leaves has been reported. Copper, zinc, boron and manganese occur in traces. Phytate phosphorus accounts for 18 percent of the total. Sulphur is required together with nitrogen for microbial protein synthesis in the rumen. Concentrations of sulphur greater than 1.5 g/kg DM or nitrogen: sulphur ratios less than 15:1 are considered adequate. Both these requirements are met in mulberry leaves.

Similarly the levels of potassium and iron in mulberry leaves are also higher than their recommended levels in diets. Higher Ca content in mulberry leaves than the required level in diet could be useful for high yielding ruminants during the early stages of lactation. Ca is closely associated with phosphorus metabolism. A high ratio of Ca:P in mulberry leaves could create some problem with calcium, phosphorus and vitamin D metabolism at a high level of supplementation of leaves in diets. It has been suggested that high Ca:P is associated with infertility in cattle. It should be noted that mulberry leaves would usually be used only as a part of livestock diets.

Feeding experiments with sheep showed that the leaves are highly palatable. The digestibility coefficients for CP, EE, CF, NFE and total carbohydrates were found to be 71, 4, 54, 84 and 76 percent respectively. The digestible nutrients per 100 kg of leaves on a DM basis were 10.7 kg CP, 0.27 kg EE, 8.3 kg CF, 40.2 kg NFE and 59.6 kg total nutrients. Mulberry leaves are also useful as cattle fodder—they are nutritious and palatable, and are stated to improve milk yield when fed to dairy animals. The feeding value of mulberry leaves is rated high by livestock owners. Feeding experiments have shown that up to 6 kg of leaves per day can be fed to milch cows without adversely affecting the health of animals or the yield and butter content of milk.

Mulberry leaf stalks, left after feeding silkworms, can also be used for feeding cattle without any adverse effect on their health and performance. The chemical composition of leaf stalks was: 11.5 CP, 7.0 true protein, 2.7 EE, 34.0 CF, 42.5 NFE, 76.5 total carbohydrates, 9.3 total ash, 1.56 Ca, 0.20 P. Digestibility coefficients of organic matter (OM), CP, EE, CF, NFE and total carbohydrate were 58, 69, 73, 49, 60 and 56 percent respectively. The balances of N, Ca and P were positive and the adult bullocks gained body weight during the experimental period.

Mulberry leaves can also be used in poultry rations. Incorporation of shade-dried mulberry leaves in layers' mash to the extent of 6 percent showed an increase in egg production with desirable yolk colour without any adverse effect on body weight and egg quality. Mulberry leaves, owing to their high carotene content, can form a valuable source of vitamin A for the health of poultry birds and increased egg production.

The effect of supplementing mulberry leaves ad libitum to concentrate diets of Angora rabbits on wool production has been studied by Singh, Goel and Negi. The average intake of mulberry leaves was 10.4 g/day/kg W 0.75 while the total DM intake was 29.5 g/day/kg W 0.75. The digestibility coefficients for DM, CP, CF and NFE were 69, 66, 72 and 78 percent respectively. The nutritive value of mulberry leaves calculated by difference was digestible CP 9.8 and total digestible nutrients 64. The results indicated that mulberry leaves can be advantageously incorporated in the diets of Angora rabbits for wool production. Mulberry leaves may be supplemented up to a level of 40 percent of the DM with impunity.

Studies have recently conducted at the Regional Station of the Indian Veterinary Research Institute on the characterisation of mulberry leaves for digestion kinetics parameters and comparison of these values with other tree foliages. Degradation kinetic parameters as studied by the *in vitro* gas production technique showed that the potential gas production in young leaves was 60.6 ml/200 mg while the rate of degradation was 0.0703.

The corresponding values for the mature leaves were 35.4 ml and 0.0624 respectively, indicating the fall in fermentability with maturity. The potential gas production for the young leaves was highest among the forages studied and the rate of gas production lower compared with only *Moringa oleifera* (Table 1), suggesting high nutritive value of the young leaves. The fermentability of the mature leaves was also high and comparable with *Leucaena* leaves. The high rate of gas production for mulberry indicates high intake potential of this forage.

Table 1

Year	*Area under mulberry (Million ha)*
1950-51	0.05673
1960-61	0.08295
1970-71	0.09424
1980-81	0.17000
1990-91	0.31310
1995-96	0.35000

The rate, potential extent and effective degradability of DM using the *in sacco* method of Orskov and McDonald were 0.0672, 85 percent and 52 percent respectively. These values for CP were 0.0467, 95 percent and 57 percent and for NDF 0.0368, 82 percent and 43 percent respectively. The effective DM degradability of various *Leucaena* species has been reported to be 46-51 percent and the value obtained in India was 51 percent (Table 3). For mulberry leaves, the kinetic parameters for CP were also comparable to those for other good quality tree fodders (Table 4).

Table 2. Potential and rate of gas production from some tree forages

Forage	Gas production	
	Potential (ml/200 mg)	Rate
Acacia catechu	24.4	0.0456
Albizzia stipulata	17.4	0.0586
Artocarpus lakoocha	51.1	0.0379
Bauhinia variegata	23.1	0.0308
Dendrocalamus hamiltonii	33.8	0.0256
Ficus roxburghii	42.6	0.0462
Leucaena leucocephala	37.2	0.0578
Morus alba (young)	60.6	0.0703
Morus alba (mature)	35.4	0.0624
Moringa oleifera	49.5	0.0852

In sacco 48-hour degradability values for DM, CP, NDF and NDF-linked N were 76, 87, 70 and 79 respectively. It is clear from these values that nitrogen linked to NDF is degraded in the rumen to a considerable extent but depending on the rate of passage could well serve as source of undegradable nitrogen. The leaves of *Artocarpus* and *Ficus* are regarded as the most valuable fodder by farmers in the hilly areas of Nepal and India.

Table 3. Rate and potential extent of dry matter degradation (PD) and effective degradability (ED) of dry matter (passage rate of 0.05/hours)

Forage	*Rate*	*PD*	*ED*
Acacia catechu	0.0390	55	33
Albizzia stipulata	0.0348	32	24
Artocarpus lakoocha	0.0953	85	57
Bauhinia variegata	0.0489	42	27
Dendrocalamus hamiltonii	0.0347	56	26
Ficus roxburghii	0.0647	93	56
Leucaena leucocephala	0.0693	75	51
Morus alba (mature)	0.0467	85	52

Table 4. Rate and potential extent of crude protein degradation (PD) and effective degradability (ED) of crude protein (passage rate 0.05/hours)

Forage	*Rate*	*PD*	*ED*
Acacia catechu	0.0295	42	14
Albizzia stipulata	1.0907	22	21
Artocarpus lakoocha	0.0693	93	64
Bauhinia variegata	1.2233	53	51
Dendrocalamus hamiltonii	0.0283	75	35
Ficus roxburghii	0.0540	95	60
Leucaena leucocephala	0.0426	78	43
Morus alba (mature)	0.0672	95	57

The degradability values for mulberry are similar to those for *Leucaena leucocephala, Artocarpus lakoocha* and *Ficus roxburghii* (Table 5). The contents of NDIN and ADIN in mulberry leaves were 56.5 and 20.4 percent of the total nitrogen respectively. The solubility of nitrogen in borate phosphate buffer was 17.3 percent and in phosphate buffer 15.7 percent. These values are of order similar to those for *Leucaena, Artocarpus, Dendrocalamus* and *Ficus* (Table 6). It was interesting that all the soluble N in mulberry was in the NPN form.

Table 5. In sacco degradability (48 hours) of dry matter (DM), crude protein (CP), neutral detergent fibre (NDF) and neutral detergent fibre bound nitrogen (NDF-N)

Forage	*DM*	*CP*	*NDF*	*NDF-N*
Acacia catechu	47	34	28	11
Albizzia stipulata	30	25	2	0
Artocarpus lakoocha	85	92	75	90
Bauhinia variegata	42	64	20	27
Dendrocalamus hamiltonii	45	57	39	51
Ficus roxburghii	88	86	84	87
Leucaena leucocephala	72	71	58	56
Morus alba (mature)	76	87	70	79

Table 6. Solubility of nitrogen (N) for some tree leaves

Forage	*(% of total N)*	
	Borate phosphate buffer (pH 8.1)	*Phosphate buffer (pH 6.8)*
Acacia catechu	21.5	13.0
Albizzia stipulata	21.6	12.2
Artocarpus lakoocha	16.9	14.6
Bauhinia variegata	42.0	23.5
Dendrocalamus hamiltonii	14.5	12.3
Ficus roxburghii	14.4	14.8
Leucaena leucocephala	19.1	13.3
Morus alba (mature)	17.3	15.7

From the *in sacco* data, the values for rumen degradable nitrogen (RDN), rumen undegradable nitrogen (UDN) and rumen undegradable but available post ruminally (DUN) calculated are shown in Table 7, and the values for digestible organic matter, metabolizable energy and intake potential calculated from the gas production data in Table 8.

Table 7. Rumen degradable nitrogen (RDN), rumen undegradable nitrogen (UDN) and rumen undegradable but available post ruminally (DUN) of some tree forages calculated at a passage rate of 0.05/hours

Forage	*RDN*	*UDN*	*DUN*
Acacia catechu	4.8	28.7	16.0
Albizzia stipulata	7.3	28.4	11.6
Artocarpus lakoocha	19.7	11.4	4.6
Bauhinia variegata	16.9	16.1	9.45
Dendrocalamus hamiltonii	11.1	20.6	11.2
Ficus roxburghii	13.2	8.9	2.6
Leucaena leucocephala	14.6	19.1	5.8
Morus alba (mature)	16.3	12.2	5.8

It is evident from the data presented in these tables that the nutritive value of mulberry is as high as some well-known good quality fodders. In addition, tannins at high levels are present in the leaves of most trees, whereas mulberry leaves are free of tannins. Mulberry leaves have the potential to be used as a supplementary feed for increasing livestock productivity in crop residue-based livestock systems. It has been a popular misconception that the low productivity of ruminants in developing countries is mainly the result of the low energy density of available forages.

There is now abundant evidence that low productivity stems from an inefficient utilisation of the feed resources because of deficiencies of nutrients in the diet. Maximum fermentation rates are attained when all factors required by the ruminal micro-organisms are available—a source of energy, nitrogen, sulphur and minerals, and also when the nitrogen release and the energy availability are synchronised. Ruminants fed low-quality forages require supplementation with the critically deficient nutrients to optimise productivity. Since mulberry leaves are rich in nitrogen, sulphur and minerals their supplementation could increase the efficiency of utilisation of crop residues by increasing the efficiency of microbial protein synthesis in the rumen, leading to higher microbial protein supply to the intestine.

Table 8. Digestible organic matter (DOM), metabolisable energy (ME) and intake potential calculated from the gas data production data

Forage	DOM (%)	ME (MJ/kg)	Intake potential (g/kg W0.75)
Acacia catechu	45	10.5	58.2
Albizzia stipulata	44	6.8	65.8
Artocarpus lakoocha	63	9.1	58.8
Bauhinia variegata	42	7.6	47.7
Dendrocalamus hamiltonii	40	6.7	46.5
Ficus roxburghii	54.9	9.2	62.0
Leucaena leucocephala	59.1	13.1	69.6
Morus alba (mature)	64	11.3	61-72
Moringa oleifera	74	9.5	-

Recent concepts of diet formulation are based on the manipulation of the diet in order to achieve high microbial efficiency and high production of microbial protein in the rumen by creating an efficient rumen ecosystem. An associated advantage of achieving high efficiency of microbial protein synthesis in the rumen is the lower emission of environmental polluting gases—methane and carbon dioxide, from feeds. The combination of trees and grassland would obviously be both a desirable development and synergistic for cattle production.

Strategic supplementation is justified because of regular feed shortages that occur and the fact that ruminants subsist for most of their life on fibrous crop residues on small farms. For high producing animals the feeding strategy should also be aimed at supplementation with bypass protein of dietary origin in addition to creating an efficient ecosystem. The genetic potential of superior livestock will only be realised if feeding systems are developed to ensure that the supply of essential dietary nutrients matches the requirements for higher levels of production.

The use of high protein mulberry tree fodder should be encouraged to provide supplementation of crop residues and natural pastures, thereby increasing productivity and the overall use of available on-farm biomass. Mulberry tree forage is well accepted by ruminants, pigs, poultry and rabbits. There is a need for systematic research on the optimisation of the use of this tree forage and for developing strategies for its optimum supplementation under different feeding situations. The promotion of mulberry should be viewed in context of a holistic farming systems approach with the aim to increasing farmers' incomes, generating employment and conserving the environment. An attractive option to achieve an integration of silk production and livestock rearing. Acceptance of these strategies could reduce the need for land clearing and pasture establishment in the fragile areas of the world that are prone to erosion following clearing.

According to one estimate, a total of approximately 1.2 billion hectares of land in the world is degraded, barren or marginal and this proportion is increasing every year. Mulberry is an ideal tree species for economic management of unutilised wasteland for the following reasons:

- it is tap rooted with minimum superficial roots;
- it has good coppicing power and is tolerant to lopping and pruning;

- pruning and training of mulberry enhances the size and quality of its leaves;
- it has easy generation capacity through seeds and vegetative means;
- it is a multipurpose tree that yields fodder, fibre, fruit, wood, etc. The leaves are highly palatable and nutritious for livestock, yet they are used extensively for silk production;
- many varieties of mulberry are available that can grow in varied agroclimatic conditions in both temperate and tropical areas.

Mulberry has the potential to play a valuable role in world agriculture. It is an extremely versatile plant that can fulfil a number of roles in smallholder agricultural production. Its value is multifaceted and the potential for increasing and diversifying its use is enormous. However, its value and benefits as a high-quality supplement to low-quality roughages in ruminant feeding systems have not been widely known nor fully exploited. There is a wide genetic diversity in this species which has wide-ranging soil and climatic adaptation—a large number of provenances are available that grow under different soil and climatic conditions. Systematic studies are warranted to evaluate these provenances in order to know the superior genotypes, collect and maintain germplasm, and conduct agronomy and management studies.

9

Integration of Mulberry Cultivation

Pond mud or humus is the main nutrient source for mulberry cultivation. After fish harvest in winter each year, the pond mud is removed to form the "base soil" (base manure) of mulberry plants. The soil added is in a solid and dry form known as adding "Tani" in Kwantong Province, and "Baini" in Zhejiang Province. During summer and autumn, liquid humus known as "Nihua" or "Hsiaoni" are applied 2-3 times in smaller quantity. At the end of each year, the base soil of the mulberry plants has increased by 5-6 cm thick.

According to the estimation of the farmers from Shunde County, the fertilisation efficiency of 100 tons of "liquid humus" is equivalent to that of 50 kg of chemical fertilisers. The amount of pond mud that are added on the base of the mulberry plants at the end of the year is estimated to be approximately equivalent to the application of 15 kg of urea. The pond mud which are applied on the base of the mulberry plants not only acts as a source of fertiliser but also resists drought and prevents the growth of weeds aside from having a longer effective time of fertilisation; especially during high temperature in summer and drought, addition of pond humus has increased significantly the production of mulberry leaves.

Due to the variation in climatical and other natural conditions, integrated fish farms in the Pearl River delta produce 2,400–2,500 kg of mulberry leaves per mu sufficient to rear 8 cycles of silkworm whilst 1000–1500 kg of mulberry leaves per mu are produced in the Taihu basin rearing 4–5 cycles of silkworms. Fifty kg of mulberry leaves can produce 4 kg of silkworm cocoons. In Pearl River delta, production of cocoons from the one mu of mulberry field is between 160 and 200 kg. In Taihu basin, the production is 80–120 kg/mu.

For every 50 kg of cocoons, an average of 3.5 kg of raw silk can be obtained. The average production of raw silk per mu of mulberry plants in Pearl River delta is 11.2–14 kg, while in Taihu basin, the production is 5.6–8.4 kg. After processing, the silk are exported and will not be recycled back to the system. Other organic waste or subsidiary products such as pupae, worm dregs and the waste silk extraction from boiled cocoons again enter the process of recycling.

According to a survey, 50 kg of mulberry leaves can produce 4 kg of cocoons with 15 kg (in Taihu basin) or 25 kg of worm dregs which are comprised of the feces and sloughs of silkworm as well as mulberry residues. These worm dregs can be used to fertilise fish ponds or as feedstuff of fish. Usually 4 kg of worm dregs produce 0.5 kg of fish. Each mu of mulberry plants yields 300–450 kg of worm dregs in Taihu basin and 1,200–1,250 kg in Pearl River delta.

By just feeding the pond fish with worm dregs alone, each mu of mulberry can produce 37.5–56.2 kg of fresh fish in Taihu basin and 150–156.25 kg in Pearl River delta. The production of cocoons from per mu of mulberry also varies with locality. However, the amount of waste water obtained after extraction of silk from the cocoons is about

2,500 kg per mu irrespective of locality. Using the experience of fish farms and their production output in Wuxi as basis, generally, 100 kg of waste water from silk extraction of cocoons can raise 0.5 kg of fish. Hence, just waste water alone can produce 12.5 kg of fresh fish.

The convertion rate of pupae to fish is 1.5–2.0:1. Using 2 as the convertion ratio, 50 kg of cocoons can yield 40 kg of pupae. Thus one mu of mulberry after raising cocoons can yield 64–96 kg of pupae in Taihu basin and 128–140 kg in Pearl River delta. If all the pupae are solely used as fish feeds, each mu of mulberry can yield 32–48 kg of fresh fish in Taihu basin and 64–70 kg in Pearl River delta (Table 1).

Table 1. The basis of integration of mulberry cultivation-sericulture and fish farming model in Taihu basin and Pearl River delta.

Items	*Jinwand Production Bridge*	*Taihu Basin*	*Pearl River Delta*
mulberry leaves (kg/mu)	875	1000–1500	2400–2500
cocoons (kg)	44.88	80–120	160–200
wormdregs (kg)	354.15	300–450	1200–1250
	(44.25)	(37.50–56.25)	(150–156.2)
pupae (kg)	35.90	64.96	128–140
	(17.95)	(32–48)	(64–70)
waste water from silk extraction (kg) 1795	2500	2500	
	(8.95)	(12.5)	(12.5)
raw silk (kg)	5.85	5.6–8.4	11.2–14
total fish produced (kg)	71.20	82–116.75	226.5–238.75

Thus, each mu of mulberry can provide the basic requirements of feeds and fertilisers to yield 82–116.75 kg of fish in the Taihu basin and 226.5– 238.75 kg in the Pearl River delta. Of course, parts of the feedstuff that are given to the fish in the pond may not be eaten but deposited on the pond bottom and together with the pond mud are recycle to manure the mulberry plants.

Apparently, the area for mulberry cultivation and fish farming must be in harmonious proportion to ensure sufficient supply of fertilisers and feeds for fish. According to the experience of farmers from Pearl River delta, each mu of mulberry plants can yield worm dregs, pupae and waste water to provide feeds and fertilisers sufficient for 1 mu of fish ponds. Hence, the ratio of areas of dike and water surface is 1:1. However, if there are other additional source of feeds or fertilisers, then the ratio of water surface and dikes can be reduced to 6:4 or 7:3. However, in actual practice, the ratio of water area to dike surface area is more complicated and generally, governed by traditional practice.

The Dongson Commune of Wuxian County in Suchow, for example, still retains a traditional type "Mulberry plot-fish pond". A 6-mu (0.4 ha) fish pond with 3.6 mu (0.24 ha) water area and 2.4 mu (0.16 ha) dike surface area. The water/ dike ratio is 6:4. Some of the farmers are still using 6 mu as the basic unit for their calculations of production output and its value. In relatively aged "mulberry plot-fish ponds", the dikes near the pond gate tend to collapse due to weathering by heavy rain and heat; each side of the dike tend to become loss steep as a result of erosion. If such ponds are not regularly repaired, the area for mulberry cultivation will certainly be reduced each year.

Mulberry Plot-Fish Ponds

"Mulberry plot fish ponds comprises two semi-closed manmade ecological systems, one of which, as described earlier, is formed by the integration of mulberry cultivation, silkworm rearing and fish farming into a terrestial-aquatic ecosystem known as the larger ecological system. Within this large system, exists a semi-closed aquatic ecosystem formed by the fish pond. Mulberry forms the base of the larger ecosystem while fish pond constitutes the subject component. The fish pond provides the basic requirements for mulberry cultivation and pond mud for dike maintenance. The pond itself is the base for inputs of feedstuff and fertilisers.

Sufficient supply of feedstuff and fertilisers can sustain a higher fish biomass and hence a larger quantity of fish excrement as well as a thicker layer of pond mud, serving as an excellent base manure for growth. The mulberry leaves are the main but indirect source of nutrients for fish production. Hence, analysing the system as a whole, maintaining a well developed growth base of the integrated farming system, i.e. mulberry cultivation, is the fundamental step to ensure efficient energy recycling in a "mulberry plot-fish pond". Mulberry cultivation, silkworm rearing and fish farming complement each other in the man-made ecosystem.

However, irrespective of whether in Pearl River delta or Taihu basin, such ecosystem is rather incomplete. Most of them have not incorporated silk reeling industry. If silk reeling factory is established within the "mulberry plot-fish pond" ecosystem and pupae, worm dregs and waste water from the factory can be adequately and effectively channelled into the ponds, thus enabling a more complete, recycling of organic waste. The establishment of the factory reduces transportation cost and instead increase the rate of economic returns. This is the real type of "mulberry plot-fish pond" integrated fish farming model.

The actual yield from a 6 mu "mulberry plot-fish pond" belonging to Jinwan Brigade is shown in Table 2. The actual produce from the farm was 107.7 kg of cocoons, valued at 330 yuan; worm dregs as fertilisers for producing woffia which produced 20,400 summer fingerlings (5 cm carps) valued at 57.12 yuan (28 yuan/ 10,000 fingerlings). The total value of both products were 387.12 yuan.

Table 2. Yield from a 6-mu mulberry plot-fish pond at Jinwan Production Brigade.

Item	*Yield (kg)*	*Average yield per MU*	*Remakrs (kg)*
mulberry leaves only.	2100	8750	used for feeding silkworm
cocoon	107.7	44.9	4 rearing cycles of silkworm.
worm dregs	850	354	for the yield of 2550 kg of wolffia which produces 1.5 cm - 20400 grass carp fingerlings.
fish yield	2171.5	338	net yield from 3.6 mu water area.

The cost of producing 1217.5 kg of food fish is shown in Table 3. Among the feedstuff and fertilisers, 282 man-days have been spent to collect 70,500 kg of aquatic plants from Taihu lake for feeding the fish, in addition, 150 kg of rape-seed cake, 2,000 kg of pig and sheep manure, 500 kg of nightsol have to be bought. Apparently, there is no close coordination between mulberry cultivation, silkworm rearing and fish farming. The only close integration is the use of pond humus/ mud – the base manure for mulberry cultivation. As it is the production cost per kilogram of fish is 1.216 yuan.

Table 3. Feeds, fertilisers & production cost of a 6-mu mulberry plot-fish pond (3.6 mu water area) at Jinwan Production Brigade.

Item	Weight (kg)	Average Price	Cost	Remarks
fingerlings (3590 ind.)	181.5		416.25	
aquatic grass	70,500	0.4	564	
pig-sheep excreta	2,000	0.5	20	
nightsoil	500	0.2	2	
rapeseed cake	150	10.5	31.5	
disease treatment (medication)			10	
electricity			86.8	31 hours 2.8 yuan/hr.
equipment depreciation			100	
labour cost in both stocking & harvesting, etc.			30	15 man-days at 2 yuan/ per man-day
total			1,260.55	

If Jinwan Production Brigade is able to freely utilise the mulberry residues and worm dregs and waste water to raise food fishes by constructing a silk reeling factory, then 107.7 kg of cocoons will yield 14 kg of raw silk and the side products of 86.15 kg of pupae, 4308 kg of waste water and 850 kg of worm dregs can be used to produce 170.8 kg of fish. This supplies 14% of the feeds and fertilisers inputs. Besides the scrappy silk obtained from silk reeling after further processing can be used for making blankets, cushins and pillows.

The scrappy silk together with silkworm moulds and sericin make up 7% of the total cocoon weight. From the economic stand point, the production of raw silk compared with the sale of cocoons alone will increase income by about 29.4%. Thus the establishment of a mulberry cultivation-silkworm rearing-silk reeling-fish farming integrated farming system not only enable a closer integration of the recycling of matter within the "mulberry plot-fish pond" ecosystem, but also creating a more compact and complete energy flow whereby the dikes of the pond are used for mulberry cultivation; the mulberry leaves for silkworm culture; cocoons for silk reeling and its by-products consisting of worm dregs, pupae and organic waste water for fish farming.

From socio-economic view point, it is essential to economise on fish farming expenditure so as to increase the income from mulberry cultivation and silkworm rearing, creating new production industry and increase employment opportunity. At present, some of the newly-established production bases maintain dike-water surface ratio at 1:1 and depth 3.5 m but decrease to 2.5 m during fast growth summer season, leaving 1.5 m of slope between the surface of the water and that of dike which is about 0.1 mu in every mu of water surface, hence permitting cultivation of green fodders.

Table 4. Calculated fish production converted from waste of mulberry and sericulture from a 2.4-mu dikes at Jinwan fish ponds.

Item	*Production (kg)*	*Convertion rate*	*Fish production (kg)*	*Remarks*
worm dregs	850	8	106.25	
pupae	26.15	1.5–2	43.10	based on 2 as convertion rate
waste water from silk extraction	4308	200	21.5	
Total			170.85	

In actual fact, the average water depth is about 1.5 meter before June, therefore, leaving about 1/5 of the slope of dike in the first half of the year and 1/10 in the second half. Base on the production rate of 5,000 kg of green fodder per mu, the slope can produce 750 kg of English rye grass and other fodder crops. If convertion rate of fodder crop is 30:1, then, about 25 kg of grass carp and 15 kg of silver and bighead carps can be harvested (faces produced by a 0.5 kg grass carp can fertilise the water allowing an additional cropping of 0.3 kg of silver and bighead carps).

The total production is 40 kg, and with additional fertilisers from mulberry residue and worm dregs etc, "the mulberry plot-fish pond" in Taihu basin can produce 122–156.75 kg of fish while that in Pearl River delta produce 266–278.75 kg. Vegetable and other fast-growing crops are also being cultivated on the dikes of fish ponds in Taihu basin or Pearl River delta. About 3,000 kg of vegetable were produced in each mu of dike surface in the Pearl River delta. Thus it is not difficult to achieve fish production at a level of 150–200 kg/mu in Taihu basin and 250–300 kg/mu in Pearl River delta in such "mulberry plot-fish pond" model.

If fish production is to be raised, e.g. 50 kg or higher per mu using the above model, then then there is a necessity to increase energy input. If more fish are reared, more excrement are produced; with increase in feedstuff and fertilisation, more pond mud will be accumulated, thus not only a higher fish yield will be obtained, there is also increased provision of pond mud for mulberry cultivation.

More energy input not only increase fish yield, but also increase the production potentials of the "mulberry plot-fish pond" ecosystem. Thus in the "mulberry plot-fish pond" system, pond is the most important component, the control of its production processes

Table 5. *Stocking of fingerlings of a "mulberry plot-fish pond" with expected yield of 200 kg/mu in the Taihu Lake basin.*

Fish species	Stocking			Survival rate	Harvest			
	Size (Ind/kg)	No. of Fingerlings	Weight (kg)		Weight Increment	Size (kg/Ind)	No. of fish	Net Production (kg)
Silver carp	20	130	6.50	90	13.46	0.75	117	87.5
Bighead	20	30	1.50	90	13.33	0.75	30	20
Grass carp	2	20	10.0	90	3.15	1.75	18	31.5
Common carp	40	50	1.25	85	20.00	0.6	42	25
Crucian carp	100	130	1.30	95	19.23	0.2	125	25
Total		360	20.50				332	189

regulate other land-based production activities and hence increase the production efficiency of the "mulberry plot-fish pond" ecosystem. Based on the results of the survey, a hypothetical model is suggested for the stocking of a "mulberry plot-fish pond" with an expected yield of 200 kg/mu. (Table 5).

10

Mulberry Cultivation in China

There are more than 1,000 cultivated varieties of mulberry in China. Most of them originate from the four main species: Lu mulberry (*M. multicaulis* Perr.), White mulberry (*M. alba* L.), Mountain mulberry (*M. bombycis* Koidz) and Guangdong mulberry (*M. atropurpurea* Roxb.). The main cultivated varieties of mulberry in China are the following.

Tong Xiang Qing

Indigenous and widely distributed in Zhejiang Province. It belongs to the species of Lu mulberry (*M. multicaulis* Perr.). Chromosome number is $2n = 2x = 28$.

Morphology: The tree is straight. The twigs are sturdy, long and straight, greenish-grey and somewhat yellow in colour. Internodes are straight. Internodal distance is 4.2 cm. Leaf order is 2/5. Lenticels are small and round, greyish-yellow, 8/cm^2. Winter bud is long triangle, large size, yellowish brown, adhering to the branch. Accessory buds are abundant and large. Leaf ovate, dark green. Leaf tip is sharp. Leaf margin is papillately dentate. Leaf base retuse. Leaf length is 22.8 cm, width 18.3 cm. Leaf blade is thick. Leaf surface smooth without shrivels, with good lustre. Leaves hang slightly downwards. Petiole is long and sturdy. Flowering type is monoecious. Inflorescences of staminate flowers are few, medium long.

Climatic requirement: Bud sprouting in Hangzhou is from 28 March to 6 April. Leaf-opening stage is in the middle of April. Bud sprouting ratio is 61.5 percent and the bud-growing ratio is 11.1 percent. Leaves mature in early May, belonging to the medium sprouting and mid-mature variety. Autumn leaves harden in mid-September.

Average yield per ha: This variety has medium shooting ability and few side branches. From 1 m shoot, 139 g leaves can be produced in spring and 151 g produced in autumn. There are 260 leaves/kg in spring and 200/kg in autumn. Leaves are 45.4 percent of the total weight of twigs, shoots, leaves and fruits. Annual leaf yield is 34 500 kg/ha.

Fruiting habits: Few soroses, medium-large size and purplish-black.

Leaf chemical composition: Leaf quality is good. Leaf dry matter (DM) contains 22.0-24.5 percent CP and 15.1 percent SC.

Main diseases/pests and tolerance: This variety has resistance to Shrinking-type mulberry dwarf caused by mycoplasma-like organisms (MLOs) and Brown spot disease caused by the fungi of *Septogloeum mor Briosi et Cavara*. However, it is sensitive to Black wilt bacterial disease caused by *Bacterium mori Boyer et lamb Smith* and Mulberry powdery mildew caused by *Phyllactinia corylea* (Pers) Karst.

Fertilisation recommendations: Sufficient water supply and fertilisation in summer and autumn are needed for preventing leaves from early hardening.

Hong Cang Sang

Selected from the varieties of Tong Xiang Qing, belonging to the species of Lu mulberry (*M. multicaulis* Perr.). Chromosome number is 2n = 2x = 28. Widely distributed in Zhejiang and Jiangsu Provinces.

Morphology: The canopy is somewhat spreading. The twigs are sturdy, long and straight and purplish-brown in colour. Internodes are comparatively straight. Internodal distance is 4.0 cm. Leaf order is 2/5 or 3/8. Lenticels are nearly round or elliptical, brown, medium size, 8/cm^2. Winter bud is triangle, large, light purplish-brown, slanting outwards. Accessory buds are large but few. Leaf ovate, dark green. Leaf tip is acute. Leaf margin is papillately dentate. Leaf base retuse. Leaf length is 21 cm, width 15.5 cm. Leaf blade is thick. Leaf surface smooth with good lustre. Leaves hang slightly downwards. Petiole is a long and sturdy. Flowering type is monoecious, bearing only or both staminate and pistillate flowers in a separate cluster. Inflorescences of staminate flowers are few, medium long.

Climatic requirement: Bud sprouting in Hangzhou is from 30 March to 6 April. Leaf-opening stage is from 7 to 18 April. Bud sprouting ratio is 55 percent and the bud-growing ratio is 19.8 percent. Leaves mature in early May, belonging to the medium sprouting and mid-mature variety. Autumn leaves harden in mid-September.

Average yield per ha: This variety has medium shooting ability and few side branches. From 1 m shoot, 135 g leaves can be produced in spring and 137 g produced in autumn. There are 263 leaves/kg in spring and 137/kg in autumn. Leaves are 46 percent of the total weight of twigs, shoots, leaves and fruit. Annual leaf yield is 33 900 kg/ha.

Fruiting habits: Few soroses, large size, purplish-black.

Leaf chemical composition: Leaf quality is good. Leaf DM contains 20.6-23.8 percent of CP and 16.9 percent of SC.

Main diseases/pests and tolerance: This variety has resistance to Mulberry dwarf and Brown spot disease. However, it is sensitive to the bacterial disease caused by *Pseudomonas syringae* Pv. mori Van Hall. It has stronger cold resistance than that of Tong Xiang Qing. Recommended training type is low or medium trunk with higher planting density.

Hu Sang 197

Selected from Zhejiang Province, belonging to the species of Lu mulberry (*M. multicaulis* Perr.). Chromosome number is $2n = 2x = 28$.

Morphology: Canopy is spreading. Twigs are sturdy, long and straight, light purplish-brown. Internodes are slightly curved. Internodal distance is 3.8 cm. Leaf order is 2/5. Lenticels are small and elliptical, greyish-brown, $13/cm^2$. Winter bud is a long triangle, greyish-brown, adhering to the branch. Accessory buds are few and smaller. Leaf is long-cordate, slightly tilting to one side, dark green in colour. Leaf tip is short-caudate. Leaf margin is papillately dentate. Leaf base emarginate. Leaf length is 24.5 cm, width 19.9 cm. Leaf blade is medium thick. Leaf surface smooth slightly shrivelled, good lustre. Leaves hang slightly downwards. Petiole is long and sturdy. Flowering type is pistillate.

Climatic requirement: Bud sprouting is from 30 March to 6 April. Leaf-opening stage is from 8 to 19 April. Bud sprouting ratio is 70 percent and the bud-growing ratio is 14.3 percent. Leaves mature in early May, belonging to the late-sprouting and mid-mature variety. Autumn leaves harden in late September.

Average yield per ha: This variety has strong shooting ability and the side branches are few. From 1 m shoot, 128 g leaves can be produced in spring and 170 g produced in autumn. There are 375 leaves/kg in spring and 224/kg in autumn Leaves are 47.7 percent of the total weight of twigs, shoots, leaves and fruit. Annual leaf yield is 33 750 kg/ha.

Fruiting habits: Soroses are few and small, purplish-black.

Leaf chemical composition: The leaf quality is good. Leaf DM contains 18.7-27.1 percent of CP and 13.1 percent of SC.

Main diseases/pests and tolerance: This variety is medium-resistant to Mulberry dwarf and Brown spot disease but sensitive to Black wilt bacterial disease. It has good tolerance to drought and poor fertilisation, therefore, it is recommended planting on river- and hillsides, with medium or low trunk.

Hu Sang 199

A resistant variety selected from Zhejiang, belonging to the large-leaf variety of species of White mulberry (*M. alba* L.). Chromosome number is 2n = 2x = 28. Widely cultivated in Zhejiang and Jiangsu Provinces.

Morphology: Canopy is slightly spreading. Twigs are sturdy, long and straight, purplish-brown. Internodes are comparatively straight. Internodal distance is 5 cm. Leaf order is 2/5. Lenticels are nearly round and small, yellowish-brown in colour, 12/cm^2. Winter bud is triangle, purplish-brown, adhering to the branch. Accessory buds are few and smaller. Leaf is cordate, light green. Shallowly incised leaf is occasionally found. Leaf tip is short-caudate. Leaf margin is obtuse dentate. Leaf base emarginate. Leaf length is 22.9 cm, width 21.5 cm. Leaf blade is comparatively thin. Leaf surface smooth and slightly shrivelled. Leaves hang slightly downwards. Petiole is long and sturdy. Flowering type is monoecious, bearing only or both staminate and pistillate flowers in a separate cluster. Number and length of inflorescences of staminate flowers are medium.

Climatic requirement: Bud sprouting is from 30 March to 6 April. Leaf-opening stage is from 8 to 18 April. Bud sprouting ratio is 75 percent and the bud-growing ratio is 13.4 percent. Leaves mature from 6 to 15 May, belonging to the late-sprouting and late mature variety. Autumn leaves harden in early October.

Average yield per ha: This variety has strong shooting ability. From 1 m shoot, 151 g leaves can be produced in spring and 119 g produced in autumn. There are 400 leaves/kg in spring and 239/kg in autumn. Leaves are 44 percent of the total weight of twigs, shoots, leaves and fruit. Annual leaf yield is 29 250 kg/ha.

Fruiting habits: Soroses are few but large, purplish-black.

Leaf chemical composition: The leaf quality is comparatively poor. Leaf DM contains 19.8-21.0 percent of CP and 15.3 -16.2 percent of SC.

Main diseases/pests and tolerance: This variety is resistant to Yellow-type mulberry dwarf caused by MLOs and Black wilt bacterial disease. It also has good rooting ability by cutting.

Huo Sang (Fire Mulberry)

A native variety of Zhejiang, belonging to the species of Rui-sui mulberry (*M. mizuho* Hotta.). It is a natural triploid, with chromosome number 2n = 3x = 42. Widely distributed in Zhejiang Province.

Morphology: Canopy is spreading. Twigs are sturdy, long and straight, purplish-brown in colour. Internodes are straight. Internodal distance is 4.6 cm. Leaf order is 2/5 or 3/8. Lenticels are large and round or ellipical, 13 l/cm^2. Winter bud is right triangle, reddish-brown, slanting outwards. Accessory buds are bigger but few. Leaf cordate, flatly spreading, dark green in colour. Tender leaves of this variety are red in colour. The tree looks like fire from afar. So it is called Huo Sang, the Fire-like mulberry, in Chinese. Leaf tip is long caudate. Leaf margin is acute dentate. Leaf base emarginated. Leaf length is 21.8 cm, width 21.3 cm. Leaf blade is thick. Leaf surface smooth with some ripple. Leaf lustre is good. Petiole is medium long and sturdy. Flowering type is monoecious, bearing only staminate or pistillate flowers in a separate cluster. Staminate flowers are rare.

Climatic requirement: Bud sprouting in Hangzhou is from 22 to 25 March. Leaf-opening stage is from 1 to 13 April. Bud sprouting ratio is 70 percent and the bud-growing ratio is 6 percent. Leaves mature from 16 to 26 April, belonging to the early sprouting and early mature variety. Autumn leaves harden in early and middle September.

Average yield per ha: This variety has strong shooting ability. From 1 m shoot, 131 g spring leaves or 138 g autumn leaves can be produced. There are 231 leaves/kg in spring and 139/kg in autumn. Leaves are 46.85 percent of the total weight of twigs, shoots, leaves and fruit. Annual leaf yield is 34 050 kg/ha.

Fruiting habits: Purplish-black soroses are found occasionally. The seeds do not germinate.

Leaf chemical composition: The leaf quality is medium. Leaf DM contains 21.82-23.68 percent of CP and 15.18-16.28 percent of SC.

Main diseases/pests and tolerance: This variety has resistance to Brown spot disease, but it is sensitive to Dwarf diseases, bacterial diseases, Powdery mildew disease and Dirty leaf disease (a fungal disease caused by *Clasterosporium mori* Sydow). It has medium cold resistance. Its tolerance to pruning is low, so it is recommended training as a tree type and planting in odd pieces of land, such as riversides, pondsides and around houses.

Nong Sang 8

Selected from Zhejiang Province, belonging to the species of White mulberry (*M. alba* L.). It is a diploid, with chromosome number 2n = 2x = 28.

Morphology: The tree is straight. Twigs are medium sturdy and long, somewhat curved, greenish-grey in colour. Internodal distance is 3.4 cm. Leaf order is 2/5 or 3/8. Lenticels are small, round or elliptical, 10/cm^2. Winter bud is right triangle, slanting outwards. Accessory buds are bigger and more. Leaf long cordate, dark green. Leaf tip is caudate. Leaf margin is papillately dentate. Leaf base retuse. Leaf length is 21.5 cm, width 19.9 cm. Leaf blade is comparatively thick. Leaf surface smooth and spreading. Petiole is comparatively long and sturdy. Flowering type is monoecious, bearing only staminate or pistillate flowers in a separate cluster.

Climatic requirement: Bud sprouting in Hangzhou is from 17 to 22 March. Leaf-opening stage is from 20 March to 11 April. Bud sprouting ratio is 88.4 percent and the bud-growing ratio is 14.0 percent. Leaves mature from 16 April to 21 April, belonging to the early sprouting and early mature variety. Autumn leaves harden from middle to late October.

Average yield per ha: This variety has strong shooting ability and few side twigs. From 1 m shoot, 130 g spring leaves or 163.5 g autumn leaves can be produced. There are 444 leaves/kg in spring and 250/kg in autumn. Leaves are 49.68 percent of the total weight of twigs, shoots, leaves and fruits. Annual leaf yield is 45 000 kg/ha.

Fruiting habits: Soroses are abundant, medium-large size, purplish-red.

Leaf chemical composition: The leaf quality is good. Leaf DM contains 23.45-26.13 percent of CP and 14.13-18.28 percent of SC.

Main diseases/pests and tolerance: This variety has resistance to Shrinking-type Mulberry dwarf, Brown spot disease, bacterial diseases, and Dirty leaf disease (caused by *Clasterosporium mori* Sydow). It has broad adaptability to various climatic and soil conditions. It can be propagated by cuttings and trained into a low or medium type.

Yu 2

Selected from the hybridisation of "Hu-sang 39 ~Guangdong Jing-sang", belonging to the species of White mulberry (*M. alba* L.). It is a diploid, with chromosome number $2n = 2x = 28$.

Morphology: Canopy is somewhat spreading. Twigs are medium sturdy, long and straight, greenish-grey in colour. Internodes are straight. Internodal distance is 4.6 cm. Leaf order is 2/5. Lenticels are large, round or elliptical, 8/cm^2. Winter bud is triangular or nearly round, slanting outwards. Accessory buds are larger and more abundant. Leaf cordate, flatly spreading, light green in colour. Leaf tip is sharp. Leaf margin is obtuse dentate. Leaf base retuse or linear. Leaf length is 18.8 cm, width 15.9 cm. Leaf blade is comparatively thin. Leaf surface smooth with good lustre. Leaves flatly spread or hang slightly downwards. Petiole is comparatively long and slim. Flowering type is monoecious. Inflorescences of staminate flowers are more and comparatively long.

Climatic requirement: Bud sprouting in Jiangsu is from 30 March to 8 April. Leaf-opening stage is from 10 to 20 April. Bud sprouting ratio is 78 percent and the bud-growing ratio is 20 percent. Leaves mature from 10 to 15 May, belonging to the early sprouting and mid-mature variety. Autumn leaves harden in early September.

Average yield per ha: This variety has strong shooting ability and the side twigs are less. From 1 m shoot, 130 g spring leaves or 84 g autumn leaves can be produced. There are 692 leaves/kg in spring and 278/kg in autumn. Leaves are 42 percent of the total weight of twigs, shoots, leaves and fruit. Annual leaf yield is 31 200 kg/ha.

Fruiting habits: Soroses are few, medium-large size, purplish-black.

Leaf chemical composition: The leaf quality is medium. Leaf DM contains 22.8-26.2 percent of CP and 12.7-14.5 percent of SC.

Main diseases/pests and tolerance: This variety has strong resistance to Yellow-type mulberry dwarf and resistance to Black wilt bacterial disease. However, it is easily attacked by insect pests. It has good tolerance to pruning but weak resistance to drought and cold.

Fertilisation recommendations: Higher production of leaf can be achieved by sufficient fertilisation. The recommended training is low or medium trunk. Higher density planting is suitable for shoot harvesting.

Zhong Sang 5801

Selected from the hybridisation of "Hu-sang 38 ~ Guangdong Jing-sang", belonging to the species of Guangdong mulberry (*M. atropurpurea* Roxb.). It is a diploid, with chromosome number 2n = 2x = 28.

Morphology: Canopy is somewhat spreading. Twigs are sturdy, long and straight, greenish-grey. Internodes are straight. Internodal distance is 3.7 cm. Leaf order is 2/5. Six lenticels/cm^2. Winter bud is long triangular, earth-yellow in colour, slanting outward. Accessory buds are smaller and few. Leaf long cordate or ovate, flatly spreading, light green in colour. Leaf tip is short caudate. Leaf margin is obtuse dentate. Leaf base retuse or linear. Leaf length is 17.7 cm, width 14.5 cm. Leaf blade is comparatively thick. Leaf surface smooth with lustre. Leaves flatly spread. Petiole is long and sturdy. Flowering type is pistillate.

Climatic requirement: Bud sprouting in Jiangsu starts from early April. Leaf-opening stage is from 13 to 21 April. Bud sprouting ratio is 84 percent and the bud-growing ratio is 15 percent. Leaves mature from 3 to 9 May, belonging to the medium sprouting and mid-mature variety. Autumn leaves harden in the middle of September.

Average yield per ha: This variety has strong shooting ability and few side twigs. From 1 m shoot, 134 g spring leaves or 96 g autumn leaves can be produced. There are 734 leaves/kg in spring and 266/kg in autumn. Leaves are 41 percent of the total weight of twigs, shoots, leaves and fruit. Annual leaf yield is 23 400 kg/ha.

Fruiting habits: Soroses are abundant, medium-large size, purplish-black.

Leaf chemical composition: The leaf quality is medium. Leaf DM contains 22.3-23.6 percent of CP and 11.3-13.3 percent of SC.

Main diseases/pests and tolerance: This variety has resistance to Yellow-type mulberry dwarf and Black wilt bacterial disease. But it is sensitive to Mulberry powdery mildew and the sorosis disease caused by *Ciboria carunculoids* Whetz. It has good resistance to high moisture but weak tolerance to cold. The recommended training is low or medium trunk.

Hei You Sang

Indigenous to Sichuan, China. Natural triploid (2n=3x=42), belonging to the species of White mulberry (*M. alba* L.). Mainly cultivated in south Sichuan.

Morphology: The tree is straight. Twigs are medium sturdy and long, dark brown in colour. Internodes are straight. Internodal distance is 4 cm. Leaf order is 2/5. Lenticels are elliptical, 5/cm^2. Winter bud is long triangular, brown, adhering to the branch, slightly tilting to one side. Accessory buds are few. Leaf ovate, flatly spreading, dark green in colour. Leaf tip is sharp. Leaf margin is acute dentate. Leaf base retuse or linear. Leaf length is 22 cm, width 16.5 cm. Leaf blade is comparatively thick. Leaf surface smooth with good lustre. Leaf veins are thick. Leaves hang downwards. Petiole is long and slim. Flowering type is monoecious. Pistillate flowers are more than the staminates. Most of the flowers will drop in the early stage.

Climatic requirement: Bud sprouting in Sichuan is from 17 to 22 March. Leaf-opening stage is from 28 March to 9 April. Bud sprouting ratio is 86 percent and the bud-growing ratio is 25 percent. Leaves mature from 20 to 30 April, belonging to the medium sprouting and mid-mature variety. Autumn leaves harden in early October.

Average yield per ha: This variety has medium shooting ability and the side twigs are few. From 1 m shoot, 251 g spring leaves or 137 g autumn leaves can be produced. There are 362 leaves/kg in spring and 175/kg in autumn. Leaves are 48 percent of the total weight of twigs, shoots, leaves and fruit. Annual leaf yield is 20 700 kg/ha.

Fruiting habits: Soroses are few and small, purplish-black.

Leaf chemical composition: The leaf quality is good. Leaf DM contains 22.54-30.98 percent of CP and 11.23-16.88 percent of SC.

Main diseases/pests and tolerance: This variety has medium resistance to Yellow-type mulberry dwarf, Black wilt bacterial disease, Mulberry powdery mildew and also has resistance to Dirty leaf disease and Brown spot disease caused by fungi.

Fertilisation recommendations: This variety can be propagated by cutting, layering or bud-grafting. The recommended training is low or medium trunk.

Da Hua Sang

Indigenous to Sichuan. Natural triploid (2n=3x=42), belonging to the species of White mulberry (*M. alba* L.). Mainly cultivated in south Sichuan.

Morphology: Canopy spreading. Twigs are sturdy and long, Reddish-brown in colour. Internodes are straight. Internodal distance is 4.5 cm. Leaf order is 2/5. Lenticels elliptical, 10/cm^2. Winter bud is circular cone, slanting outwards. Accessory buds are few. Leaf cordate, spreading, light green in colour. Leaf tip is sharp. Leaf margin is obtuse dentate. Leaf base retuse. Leaf length is 27.5 cm, width 23.5 cm. Leaf blade is thick. Leaf surface smooth with good lustre. Leaves hang slightly downwards. Petiole is long and sturdy. Bearing only staminate flowers. The inflorescences are more.

Climatic requirement: Bud sprouting in Sichuan is from 16 to 26 March. Leaf-opening stage is from 1 to 8 April. Bud sprouting ratio is 70.4 percent and the bud-growing ratio is 22 percent. Leaves mature in mid-May, belonging to the medium sprouting and late-mature variety. Autumn leaves harden in early October.

Average yield per ha: This variety has medium shooting ability and the side twigs are few. From 1 m shoot, 168 g spring leaves or 163 g autumn leaves can be produced. There are 408 leaves/kg in spring and 202/kg in autumn. Leaf weight occupies 46.21 percent of the total weight of twigs, shoots, leaves and fruit. Annual leaf yield is 18 000 kg/ha.

Leaf chemical composition: The leaf quality is good. Leaf DM contains 27.7-29.4 percent of CP and 11.25-12.63 percent of SC.

Main diseases/pests and tolerance: This variety has medium resistance to Yellow-type mulberry dwarf, Mulberry powdery mildew and Brown spot disease, but is lightly sensitive to Black wilt bacterial disease. The tolerance to drought and cold is comparatively weak.

Fertilisation recommendations: This variety is suitable for planting in good soil conditions so as to achieve high productivity. The recommended training is medium or high trunk.

Xiao Guan Sang

Native variety of Sichuan, China. Diploid (2n=2x=28), belonging to the species of White mulberry (*M. alba* L.). Widely distributed in the east and north of Sichuan.

Morphology: The tree is straight. Twigs are slim and long, greenish-grey in colour. Internodes are straight. Internodal distance is 3 cm. Leaf order is 2/5. Lenticels are mostly small and round, 15/cm^2. Winter bud is right triangular, light greenish-brown in colour, adhering to the branch. Accessory buds are few and smaller. Leaf ovate, flatly spreading, dark green in colour. Few incised leaves appear in the basic part of the shoot. Leaf tip is sharp. Leaf margin is papillate dentate. Leaf base emarginate. Leaf length is 18.5 cm, width 14.5 cm. Leaf blade is thick. Leaf surface smooth and slightly shrivelled, good lustre. Leaves are flatly spreading. Petiole is short and slim. Flowering type is monoecious, with short style. Inflorescences of staminate flowers are few.

Climatic requirement: Bud sprouting in Sichuan is from 12 to 15 March. Leaf-opening stage is from 17 March to 20 April. Bud sprouting ratio is 80 percent and the bud-growing ratio is 27 percent. Leaves mature in early May, belonging to the early sprouting and early mature variety. Autumn leaves harden in early October.

Average yield per ha: This variety has few side twigs. From 1 m shoot, 226 g spring leaves or 92 g autumn leaves can be produced. There are 456 leaves/kg in spring and 225/kg in autumn. Leaves are 66.7 percent of the total weight of twigs, shoots, leaves and fruit. Annual leaf yield is 18 750 kg/ha.

Fruiting habits: Soroses are few and small, purplish-black.

Leaf chemical composition: The leaf quality is medium. Leaf DM contains 18.83-21.97 percent of CP and 14.5 percent of SC.

Main diseases/pests and tolerance: This variety has medium resistance to Dirty leaf disease and Mulberry powdery mildew, but is sensitive to Yellow-type mulberry dwarf and Black wilt bacterial disease. It also easily contracts Brown spot disease. It has strong tolerance to drought and medium tolerance to cold.

Fertilisation recommendations: This variety has broad adaptability to various soil and climatic conditions. It is suitable for planting in the scattered land of hilly areas, and trained into low or medium trunk.

Jia Ling 16

A triploid breed by crossing "Xiqing" (tetraploid) with "Yu 2" (diploid). Since 1992, it has been widely cultivated in all of Sichuan and introduced to other provinces, such as Xinjiang, Guizhou, Henan and Shaanxi.

Morphology: Twigs are sturdy, straight and long, greenish-grey. Winter bud is triangular, closely adhering to the branch. Leaf order is 2/5. Internodal distance is 3.6 cm. Leaf incised, large in size, dark green in colour. Leaf tip is short and sharp. Leaf margin is acute dentate. Leaf base emarginate. Leaf length is 28 cm, width 25.5 cm. Leaf blade is thick. Leaf surface smoothly shrivelled. Petiole is short and sturdy. Bearing very few pistillate flowers, it is sterile.

Climatic requirement: Bud sprouting in Sichuan is in the middle of March, belonging to early-sprouting and mid-mature variety. Bud sprouting ratio is above 80 percent.

Average yield per ha: This variety has high productivity.

Fruiting habits: Sterile.

Leaf chemical composition: Good leaf quality. Leaf DM contains 28.05 percent of CP in average of spring and autumn.

Main diseases/pests and tolerance: This variety has relatively strong resistance to Black wilt bacterial disease. Leaf quality can be maintained for a long time in storage.

Planting recommendations: Good fertilisation and management for this variety can lead to high productivity. It is suitable for hilly and mountainous areas with high density or intercropping.

Guangdong Jing Sang

A highly heterogeneous variety originating from the Pearl River Delta of Guangdong, widely distributed in south China. Diploid (2n=2x=28), belonging to the species of Guangdong mulberry (*M. atropurpurea* Roxb.).

Morphology: Straight tree. Twigs are slim, straight and long. Bark colour is not uniform, mostly greenish-grey or brown. Internodes are straight or slightly curved. Internodal distance is 3.4 cm. Leaf order is mostly 2/5. Lenticels are round or elliptical, 8/cm^2. Winter bud is large, triangular or nearly globular, mostly slanting outwards. Ovate or cordate leaf, entire, occasionally incised or mixed. Leaf flatly spread, light green. Leaf tip is sharp or short caudate. Leaf margin is obtuse or acute dentate. Leaf base retuse or linear, not symmetrical. Leaf length is 17 cm, width 14 cm. Thin leaf blade. Smooth leaf surface not shrivelled, weak lustre. Leaves are flatly spread. Petiole is slim, medium length. Flowering type is monoecious or dioecious, occasionally both staminate and pistallate flowers are found in a separate cluster. Inflorescences of staminate flowers are abundant and long.

Climatic requirement: Bud sprouting in Guangzhou is from 18 to 21 January. Leaf-opening stage is from 2 to 21 February. Bud sprouting ratio is 85 percent and the bud-growing ratio is 20 percent. First crop leaves mature from

27 February to 17 March. Growing period is 25-30 days for whole leaf and 40-45 days for the leaf-shoot respectively, belonging to the early sprouting and early mature type.

Average yield per ha: This variety has very strong shooting ability, with many side twigs, and uniform growth. From 1 m shoot, 85 g spring leaves or 65 g autumn leaves can be produced. There are 441 leaves/kg in the main branch and 945/kg in side twigs. Annual leaf yield can reach 33 750 kg/ha.

Fruiting habits: Soroses are abundant and large, purplish-black.

Leaf chemical composition: Medium leaf quality. Leaf DM contains 21.24 percent of CP and 8.38 percent of SC.

Main diseases/pests and tolerance: This variety has resistance to Common mulberry dwarf but is sensitive to Dirty leaf disease, Mulberry powdery mildew and Green bacterial wilt (caused by *Pseudomonas solanacearum* Smith. It has strong tolerance to pruning and moisture condition, but weak resistance to drought and cold.

Planting recommendations: This variety can be propagated by seedlings and planted in at very high density (90 000-120 000 plants/ha). High yielding can be achieved by providing sufficient fertilisation after each harvest.

Lun 40

A high yielding variety selected from Guangdong Jing Sang. Natural triploid (2n=3x=42), belonging to the species of Guangdong mulberry (*M. atropurpurea* Roxb.), widely distributed in south China.

Morphology: Canopy slightly spread. Twigs are sturdy, straight and long, brown. Internodes are straight.

Internodal distance is 3.8 cm. Leaf order is 2/5 or 3/8. Lenticels are round or elliptical, 5/cm^2. Winter bud is large, oval shape, brown, slanting outwards. Accessory buds are many and large. Cordate leaf, flatly spread, light green. Tender leaves look purplish-red. Leaf tip is short caudate. Leaf margin is papillately dentate. Retuse leaf base. Leaf length is 24.5 cm, width 16.95 cm. Leaf blade is relatively thick. Smooth leaf surface not shrivelled, good lustre. Leaves hang slightly downwards. Petiole is sturdy and long. Flowering type is pistallate.

Climatic requirement: Bud sprouting in Guangzhou is from 16 to 28 January. Leaf-opening stage is from 2 to 15 February. Bud sprouting ratio is 80 percent and the bud-growing ratio is 17 percent. First crop leaves mature from 18 February to 10 March. Growing period is 25-30 days for whole leaf and 40-45 days for the leaf-shoot respectively, belonging to the early sprouting and early mature type.

Average yield per ha: This variety has medium shooting ability, with few side twigs. From 1 m shoot, 175 g spring leaves or 120 g autumn leaves can be produced. There are 178 leaves/kg in the main branch and 475/kg inside twigs. Annual leaf yield can reach 46 500 kg/ha.

Fruiting habits: Soroses are abundant and large, purplish-black, but seed fertility is very low.

Leaf chemical composition: Medium leaf quality. Leaf DM contains 21.06-25.01 percent of CP and 7-7.9 percent of SC.

Main diseases/pests and tolerance: This variety grows very fast and with strong tolerance to pruning. It is lightly sensitive to Dirty leaf disease, Green bacterial wilt caused by *Pseudomonas solanacearum* Smith and Red rust disease caused by *Accidium mori* (Board) Diet. Its resistance to drought and cold is also weak.

Planting recommendations: This variety can be planted at high density (90 000 plants/ha). Additional applications of phosphorus and potassium are needed for new plantations.

Lun 109

Selected from Pearl River Delta of Guangdong. Diploid (2n=2x=28), belonging to the species of Guangdong mulberry (*M. atropurpurea* Roxb.).

Morphology: Canopy slightly spread. Twigs are sturdy, straight and long, greyish-brown. Internodes are slightly curved. Internodal distance is 3.8 cm. Leaf order is 3/8. Lenticels are round or elliptical, 9/cm^2. Winter bud is oval shape, light brown, slanting outwards. Accessory buds are many and large. Ovate or cordate leaf, light green. Short caudate leaf tip. Leaf margin is papillately dentate, slightly bending upwards. Retuse leaf base. Leaf length is 27 cm, width 20.2 cm. Leaf blade is relatively thick. Petiole is sturdy and long. Flowering type is staminate, with long inflorescence. Pistallates are rarely found.

Climatic requirement: Bud sprouting in Guangzhou is from 16 to 31 January. Leaf-opening stage is from 5 to 21 February. Bud sprouting ratio is 68 percent and the bud-growing ratio is 27 percent. First crop leaves mature from 24 February to 10 March. Growing period is 25-30 days for whole leaf and 40-45 days for the leaf-shoot respectively, belonging to the early sprouting and mid-mature type.

Average yield per ha: This variety has strong shooting ability. It grows fast, with early sprouting and sturdy side twigs. From 1 m shoot, 147 g spring leaves or 110 g autumn leaves can be produced. There are 192 leaves/kg in the main branch and 583/kg in side twigs. Annual leaf yield can reach 39 450 kg/ha.

Leaf chemical composition: Medium leaf quality. Leaf DM contains 29.47 percent of CP and 7.61 percent of SC.

Main diseases/pests and tolerance: This variety has medium tolerance to drought but weak tolerance to cold. It is sensitive to Green bacterial wilt caused by *Pseudomonas solanacearum* Smith.

Planting recommendations: This variety can be planted at high density (90 000 plants/ha). Apply nitrogen fertiliser after each harvest and additional phosphorous and potassium after the second harvest.

Sha 2

Selected from Pearl River Delta of Guangdong. Diploid (2n=2x=28), belonging to the species of Guangdong mulberry (*M. atropurpurea* Roxb.).

Morphology: Canopy slightly spread. Twigs are sturdy, straight and long, yellowish-brown. Straight internodes. Internodal distance is 5.2 cm. Leaf order is 2/5.5 lenticels/cm^2. Winter bud is triangular, greyish-brown, adhering to the branch. Many and large accessory buds. Ovate or long cordate leaf, flatly spread, green. Tender leaves look light purplish-red. Leaf tip is short caudate. Leaf margin is papillately dentate. Linear leaf base. Leaf length is 30.7 cm, width 21.5 cm. Leaf blade is relatively thick. Petiole is sturdy and short. Flowering type is pistillate, with short style.

Climatic requirement: Bud sprouting in Guangzhou is from 20 to 31 January. Leaf-opening stage is from 3 to 23 February. Bud sprouting ratio is 62.9 percent and the bud-growing ratio is 37.1 percent. First crop leaves mature from 20 February to 10 March. Growing period is 25-30 days for whole leaf and 40-45 days for the leaf-shoot respectively, belonging to the early sprouting and early mature type.

Average yield per ha: This variety has medium shooting ability, with early sprouting and many side-twigs. From 1 m shoot, 109 g spring leaves or 81 g autumn leaves can be produced. There are 154 leaves/kg in the main branch and 294/kg in side twigs. Annual leaf yield can reach 39 000 kg/ha.

Fruiting habits: Soroses are abundant and large, purplish-black.

Leaf chemical composition: Medium leaf quality. Leaf DM contains 22.88 percent of CP and 10.88 percent of SC.

Main diseases/pests and tolerance: This variety has strong resistance to mulberry thrips (*Pseudodendrothrips mori* Niwa) and mulberry leaf mite (*Tetranychus cinnabarinus* Boisoluval and *Eotetranychus suginanamensis* Yokoyama). However, it is sensitive to Green bacterial wilt caused by *Pseudomonas solanacearum* Smith. It has tolerance to moisture and medium tolerance to drought, but its tolerance to cold is weak.

Planting recommendations: For seed production purposes, more phosphorus and potassium are needed in fertiliser. When soroses appear red, apply nitrogen fertiliser slightly to promote fruiting.

Da 10

A natural triploid (2n=3x=42) Selected from Pearl River Delta of Guangdong, belonging to the species of Guangdong mulberry (*M. atropurpurea* Roxb.).

Morphology: Canopy slightly spread. Twigs are slim, straight and long, light brown. Internodes are straight. Internodal distance is 4.8 cm. Leaf order is 1/2. Lenticels are round or elliptical, 6/cm^2. Winter bud is triangular, brown, slanting outwards. Accessory buds are many and large. Cordate leaf, spread, light green. Long caudate leaf tip. Acute dentate leaf margin. Emarginate leaf base. Leaf length is 22 cm, width 18.3 cm. Smooth leaf surface and slightly shrivelled, with weak lustre. Leaves hang slightly downwards. Petiole is sturdy and short. Flowering type is pistillate, without style.

Climatic requirement: Bud sprouting in Guangzhou is from 16 to 28 January. Leaf-opening stage is from 31 to 16 February. Bud sprouting ratio is 85 percent and the bud-growing ratio is 10 percent. First crop leaves mature from 20 February to 10 March. Growing period is 25-30 days for whole leaf and 40-45 days for the leaf-shoot respectively, belonging to the early sprouting and early mature type.

Average yield per ha: This variety has medium shooting ability, with few side twigs. Annual leaf yield is 30 000 kg/ha. Fruit production can reach 15 000 kg/ha per year.

Fruiting habits: Soroses are abundant and large, purplish-black. The fruit contains 2.78-3.8 percent of total carbohydrate, 2.28-3.57 percent of reducing sugar, 0.81 percent of amino acids. Vitamin C content of the berry reaches 1.02 mg/100 g, adequate for processing and beverage production.

Main diseases/pests and tolerance: This variety is slightly sensitive to Mulberry mosaic disease caused by virus. It is also easily attacked by insect pests. Its tolerance to cold is weak.

Planting recommendations: For fruit production purposes, more fertiliser, with phosphorous and potassium, is needed. Two prunings are possible, in April and in July/August. Nitrogen fertiliser should be applied between prunings.

Kang Qing 10

A resistant variety selected from the west of Guangdong.

Diploid (2n=2x=28), belonging to the species of Guangdong mulberry (*M. atropurpurea* Roxb.).

Morphology: Canopy slightly spread. Twigs are sturdy, straight and long, greyish-brown. The sun-facing side of the tender shoot is light reddish-brown and the top bud is curved. Internodes are straight. Internodal distance is 4.9 cm. Leaf order is 1/2. Lenticels are round, 9/cm^2. Winter bud is triangular, greyish brown, slanting outwards. Accessory buds are few and smaller. Cordate leaf, occasionally incised, spread, light green. Sharp leaf tip. Obtuse dentate leaf margin. Linear leaf base. Leaf length is 27.3 cm, width 24.4 cm. Thick leaf blade. Smooth leaf surface with good lustre. Leaves hang slightly downwards. Petiole is sturdy and short. Flowering type is staminate, with many long inflorescences. Monoecism is found occasionally.

Climatic requirement: Bud sprouting in Guangzhou is from 26 January to 5 February. Leaf-opening stage is from 11 to 24 February. Bud sprouting ratio is 80 percent and the bud-growing ratio is 15 percent. First crop leaves mature from 16 February to 5 March. Growing period is 20-25 days for whole leaf and 35-40 days for the leaf-shoot respectively, belonging to the early sprouting and early mature type.

Average yield per ha: This variety has strong shooting ability, grows fast, with many side twigs. From 1 m shoot, 130 g spring leaves or 100 g autumn leaves can be produced. There are 206 leaves/kg in the main branch and 311/kg in the side twigs. Annual leaf yield can reach 37 050 kg/ha.

Leaf chemical composition: Medium leaf quality. Leaf DM contains 22.4 percent of CP and 5.26 percent of SC.

Main diseases/pests and tolerance: This variety has strong resistance to Mulberry green bacterial wilt and

Mulberry powdery mildew, medium resistance to Mulberry red rust disease. However, it is easily attacked by mulberry thrips and mulberry leaf mite. It has weak tolerance to cold. It has strong rooting ability and can be propagated by cuttings.

Hei Lu Cai Sang

Indigenous to Shandong China, belonging to the species of Lu mulberry (*M. multicaulis* P.). Chromosome number is 2n = 2x = 28.

Morphology: Straight tree. Twigs are slim and straight, medium long. Greenish-brown bark. Internodes are slightly curved. Internodal distance is 3 cm. Leaf order is 2/5. Lenticels are elliptical, 7/cm^2. Winter bud is triangle, dark brown, adhering to the branch. Accessory buds are smaller and less. Ovate leaf, margin slightly bends upwards. Leaf colour dark green. Caudate leaf tip. Papillately dentate leaf margin. Leaf base retuse or linear. Leaf length is 18 cm, width 14 cm. Leaf blade is comparatively thick. Smooth leaf surface slightly shrivelled, with good lustre. Leaves flatly spread. Petiole is long and slim. Monoecious flowering type. Inflorescences of staminate flowers are short but abundant.

Climatic requirement: Bud sprouting in Shandong is from 22 to 25 April. Leaf-opening stage is from 28 April to 8 May. Bud sprouting ratio is 75 percent and bud-growing ratio is 15 percent. Leaves mature on 14 May, belonging to the medium sprouting and mid-mature type. Autumn leaves harden in early September..

Average yield per ha: This variety has strong shooting ability with many side branches. From 1 m shoot 98 g leaves can be produced in spring and 116 g in autumn. There are 670 leaves/kg in spring and 275/kg in autumn.

Leaves are 49.3 percent of the total weight of twigs, shoots, leaves and fruits. Annual leaf yield is 19 500 kg/ha.

Fruiting habits: Soroses are few and small, purplish-black.

Leaf chemical composition: Good leaf quality. Leaf DM contains 23.8-24.1 percent of CF and 13.2-18.3 percent of SC.

Main diseases/pests and tolerance: This variety has medium resistance to Brown spot disease. It has strong cold resistance and good tolerance to wind. *Planting recommendations*: Recommended for Yellow River area and to be trained into low or medium trunk.

Da Ji Guan Sang

Indigenous to Shandong, belonging to the species of Lu mulberry (*M. multicaulis* Perr.) Chromosome number is 2n = 2x = 28.

Morphology: Canopy is slightly spread. Twigs are sturdy, short and straight. Brown bark. Straight internodes. Internodal distance is 3 cm. Leaves have no order. Lenticels are small and round, 5/cm^2. The top of the shoot swells like a cock's comb, so it is called Da Ji Guan (Big cock's comb) in Chinese. Brownish-red winter buds. The buds in the middle and lower parts of the shoot are triangular or nearly globular, some of them are slanting outwards. Few and small accessory buds.

Ovate leaf, spread or leaf margin slightly bent upwards. Dark green leaves. Sharp leaf tip. Papillately dentate leaf margin. Retuse leaf base. Leaf length is 24 cm, width 20 cm. Thick leaf blade. Smooth leaf surface not shrivelled, with good lustre. Leaves hang slightly downwards. Petiole is sturdy and short. Monoecious flowering type. Inflorescences of staminate flowers are short but abundant.

Climatic requirement: Bud sprouting in Shandong is from 22 to 26 April. Leaf-opening stage is from 29 April to 7 May. Bud sprouting ratio is 85 percent and bud-growing ratio is 25 percent. Leaves mature on 16 May, belonging to the medium sprouting and mid-mature variety. Autumn leaves harden in mid-August.

Average yield per ha: This variety has medium shooting ability with many side branches. From 1 m shoot 212 g leaves can be produced in spring and 184 g in autumn. There are 660 leaves/kg in spring and 240/kg in autumn. Leaves are 59.3 percent of the total weight of twigs, shoots, leaves and fruit. Annual leaf yield is 20 000 kg/ha.

Fruiting habits: Soroses are few and small, purplish-black.

Leaf chemical composition: Good leaf quality. Leaf DM contains 23.1-27.2 percent of CP and 12-17.1 percent of SC.

Main diseases/pests and tolerance: This variety has strong resistance to cold and wind. It has low tolerance to pruning. It is sensitive to Black wilt bacterial disease, Mulberry red rust and Brown spot disease.

Planting recommendations: Recommended for Yellow River area and to be trained into medium or high trunk.

Xuan 792

Selected from the variants of native mulberry in Shandong, belonging to the species of Lu mulberry (*M. multicaulis* Perr.). Chromosome number is 2n = 2x = 28. Widely distributed in Shandong and provinces in North China.

Morphology: Canopy is slightly spread. Twigs are sturdy, long and straight. Brown bark. Internodes are slightly curved. Internodal distance is 3.5 cm. Leaf order is 2/5. Lenticels are round, 5/cm^2. Winter bud is triangular, adhering to the branch. Accessory buds are small and few. Ovate leaf, spread, dark green. Leaf tip is sharp or short caudate. Obtuse dentate leaf margin. Linear leaf base. Leaf length is 23 cm, width 17 cm. Thick leaf blade. Smooth leaf surface and not shrivelled, with good lustre. Leaves hang slightly downwards. Petiole is medium sturdy and long. Pistillate flowering type.

Climatic requirement: Bud sprouting in Shandong is from 24 to 28 April. Leaf-opening stage is from 1 to 10 May. Bud sprouting ratio is 73 percent and the bud-growing ratio is 15 percent. Leaves mature on 16 May, belonging to the late sprouting and mid-mature variety. Autumn leaves harden in early September.

Average yield per ha: This variety has strong shooting ability with few side branches. From 1 m shoot 112 g leaves can be produced in spring and 138 g in autumn. There are 410 leaves/kg in spring and 190/kg in autumn. Leaves are 49.79 percent of the total weight of twigs, shoots, leaves and fruits. Annual leaf yield is 30 000 kg/ha.

Fruiting habits: Soroses are few, medium size, purplish-black.

Leaf chemical composition: Good leaf quality. Leaf DM contains 25.2-27.5 percent of CP and 11.3-15.3 percent of SC.

Main diseases/pests and tolerance: This variety has medium resistance to Yellow-type mulberry dwarf, strong tolerance to wind, cold and drought. It is lightly sensitive to Shrinking-type mulberry dwarf.

Planting recommendation: This variety is suitable to plant in the north of China and to train into low or medium trunk. Higher planting density and application of fertiliser in autumn can lead to higher production.

Niu Gen Sang

Native variety of Hebei China, belonging to the species of Lu mulberry (*M. multicaulis* Perr.). Chromosome number is 2n = 2x = 28. Widely distributed in Hebei Province.

Morphology: Canopy is slightly spread. Twigs are sturdy and long, slightly curved. Greyish brown bark. Internodes are slightly curved. Internodal distance is 4.5 cm. Leaf order is 2/5. Lenticels are comparatively small, round or elliptical, 9/cm^2. Winter bud is right triangular, light yellowish-brown in colour, slanting outwards. Accessory buds are small and few. Cordate leaf, spread, dark green. Sharp leaf tip. Papillately dentate leaf margin. Retuse leaf base. Leaf length is 23.5 cm, width 19.5 cm. Leaf blade is thick. Smooth leaf surface, slightly shrivelled. Leaf margin is bent slightly upwards. Sturdy leaf vein. Petiole is sturdy and long. Flowering type is monoecious. Inflorescences of staminate flowers are few and short.

Climatic requirement: Bud sprouting in Hebei is from 6 to 12 May. Leaf-opening stage is from 14 to 20 May. Bud sprouting ratio is 72 percent and the bud-growing ratio is 20 percent. Leaves mature from 25 May to 1 June, belonging to the medium sprouting and mid-mature type. Autumn leaves harden from 10 to 15 September.

Average yield per ha: This variety has medium shooting ability, with few side branches. From 1 m shoot 155 g leaves can be produced in spring and 140 g in autumn. There are 400 leaves/kg in spring and 160/kg in autumn. Leaves are 50 percent of the total weight of twigs, shoots, leaves and fruit. Annual leaf yield is 15 000 kg/ha.

Fruiting habits: Soroses are few, medium size, purplish-black.

Leaf chemical composition: Medium leaf quality. Leaf DM contains 16.25 percent of CP and 13.41 percent of SC.

Main diseases/pests and tolerance: This variety has resistance to Mulberry anthracnose disease caused by *Colletotrichum morifolium* Hora and the Mulberry leaf blight caused by *Hormodendrum mori* Yendo. It is sensitive to Black wilt bacterial disease and Dirty leaf disease. Its tolerance to cold and drought is medium.

Planting recommendation: This variety is suitable for planting in the north of China and for training into medium trunk, high trunk or tree type. Properly high planting density is recommended.

Hong Pi Wa Sang

Native variety of Hubei China, belonging to the species of Lu mulberry (*M. multicaulis* P.). Chromosome number is 2n = 2x = 28.

Morphology: Canopy is spread. Twigs are sturdy, long and straight. Brown bark. Internodes are straight. Internodal distance is 4 cm. Leaf order is 2/5. Lenticels are round or elliptical, 8/cm^2. Winter bud is triangular, brown in colour. Buds are slanting outwards or adhering to the branch. Few accessory buds. Cordate leaf, light green. Sharp leaf tip. Papillately dentate leaf margin. Leaf base retuse or linear. Leaf length is 21 cm, width 19 cm. Thick leaf blade. Leaf surface is smooth and shrivelled, with weak lustre. Leaves hang slightly downwards. Petiole is medium sturdy and long. Pistillate flowering type.

Climatic requirement: Bud sprouting in Zhenjiang is from 13 to 17 April. Leaf-opening stage is from 18 to 23 April. Bud sprouting ratio is 56 percent and bud-growing ratio is 15 percent. Leaves mature from 15 May to 20 May, belonging to the late sprouting and late mature type. Autumn leaves harden in the middle of September.

Average yield per ha: This variety has weak shooting ability, but grows fast with few side branches. From 1 m shoot 119 g leaves can be produced in spring and 120 g in autumn. There are 548 leaves/kg in spring and 165/kg in autumn. Leaves are 54 percent of total weight of twigs, shoots, leaves and fruits. Annual leaf yield is 18 000 kg/ha.

Fruiting habits: Soroses are small and few, purplish-black.

Leaf chemical composition: Medium leaf quality. Leaf DM contains 23.6-26.4 percent of CP and 10.8-11.5 percent of SC.

Main diseases/pests and tolerance: This variety has medium resistance to Black wilt bacterial disease, Mulberry powdery mildew and Dirty leaf disease. It is sensitive to Yellow-type mulberry dwarf.

Planting recommendation: This variety is suitable for planting in the Yangtze River area, trained into medium or high trunk type. Maintaining more shoots or proper higher planting density are recommended.

Hei Ge Lu

Native variety of Shanxi China, belonging to the species of White mulberry (*M. alba* L.). Chromosome number is 2n = 2x = 28.

Morphology: Canopy is slightly spread. The twigs are medium sturdy, long and straight. Brown bark. Straight internodes. Internodal distance is 3.3 cm. Leaf order is 2/5. Lenticels are round, small and few, 8/cm^2. Winter bud is right triangular, brown, slanting outwards. Accessory buds are few and not obvious. Ovate leaf, spread, dark green. Sharp leaf tip. Obtuse dentate leaf margin. Retuse leaf base. Leaf length is 19.4 cm, width 16.8 cm. Thick leaf blade. Smooth leaf surface, slightly shrivelled, with good lustre. Leaves hang slightly downwards. Petiole is medium sturdy and long. Monoecious flowering type. Inflorescences of staminate flowers are few and short.

Climatic requirement: Bud sprouting in Shanxi is from 18 to 25 April. Leaf-opening stage is from 27 April to 4 May. Bud sprouting ratio is 80 percent and the bud-growing ratio is 27 percent. Leaves mature from 20 to 26 May, belonging to the medium sprouting and medium mature type. Autumn leaves harden in 20 September.

Average yield per ha: This variety has strong shooting ability, with few side branches. From 1 m shoot 125 g of leaves can be produced in spring and 140 g in autumn. There are 428 leaves/kg in spring and 256/kg in autumn. Leaves are 41 percent of total weight of twigs, shoots, leaves and fruit. Annual leaf yield is 20 250 kg/ha.

Fruiting habits: Soroses are small and few, purplish-black.

Leaf chemical composition: Medium leaf quality. Leaf DM contains 23.4 percent of CP and 13.2 percent of SC.

Main diseases/pests and tolerance: This variety has medium resistance to Mulberry powdery mildew. It is lightly sensitive to Yellow-type mulberry dwarf and sensitive to Black wilt bacterial disease. It has strong tolerance to drought, cold and poor soil conditions.

Planting recommendation: This variety is suitable for planting by road- and riversides, trained into medium or high trunk type.

He Tian Bai Sang

A triploid variety ($2n = 3x = 42$) selected from south

Xinjiang, belonging to the species of White mulberry (*M. alba* L.).

Morphology: Canopy is spread. Twigs are slim and long, some bending. Brown bark colour. Straight internodes. Internodal distance is 5 cm. Leaf order is 2/5. Lenticels are large, mostly elliptical, 5/cm^2. Winter bud is long triangular, brown, adhering to the branch. Some buds are tilting to one side. Accessory buds are larger and more. Cordate leaf. Entire and incised leaves are mixed together. Incised leaf has 1-10, mostly 5, gaps. Leaf is flatly spread, dark green. Short caudate leaf tip. Obtuse dentate leaf margin. Emarginate leaf base. Leaf length is 12.6 cm, width 11.6 cm. Thick leaf blade. Smooth leaf surface, not shrivelled, with weak lustre. Leaves hang slightly downwards. Petiole is slim and medium long. Pistillate flowering type.

Climatic requirement: Bud sprouting in south Xinjiang is from 13 to 16 April. Leaf-opening stage is from 17 to 22 April. Bud sprouting ratio is 82 percent and bud-growing ratio is 21 percent. Leaves mature from 10 to 15 May, belonging to the medium sprouting and medium mature type. Autumn leaves harden in early September.

Average yield per ha: This variety has strong shooting ability with many side branches. From 1 m shoot 166 g of leaves can be produced in spring and 45 g in autumn. There are 476 leaves/kg in spring and 348/kg in autumn. Leaves are 46.9 percent of total weight of twigs, shoots, leaves and fruit. Annual leaf yield is 17 700 kg/ha.

Fruiting habits: Soroses are abundant, medium size, jade-white and sweet, (containing 18.9 percent sugar).

Leaf chemical composition: Good leaf quality. Leaf DM contains 22.1-23.3 percent of CP and 11.0-14.2 percent of SC.

Main diseases/pests and tolerance: This variety has strong resistance to drought and cold.

Planting recommendation: The recommended training type is low or medium trunk. Higher production can be achieved by providing good fertilisation and irrigation. If for both fruit and leaf production, high trunk or tree type is preferable.

Yun Sang 2

A diploid variety (2n = 2x = 28) selected from Yunnan, belonging to the species of White mulberry (*M. alba* L.).

Morphology: Canopy is slightly spread. Twigs are medium sturdy, straight and long, greyish-green. Internodes are straight. Internodal distance is 4.3 cm. Leaf order is 2/5. Lenticels are round or elliptical, 7/cm^2. Winter bud is long triangular, reddish-brown, slanting outwards with slightly tilting to one side. Accessory buds are small and few. Long cordate leaf, flatly spread, light green. Leaf tip is short caudate or sharp. Obtuse dentate leaf margin. Leaf base retuse or linear. Leaf length is 19.5 cm, width 15 cm. Thin leaf blade. Smooth leaf surface, shrivelled or not, with weak lustre. Petiole is medium in size and length. Flowering type is pistillate.

Climatic requirement: Bud sprouting in Yunnan is from 1 to 21 February. Leaf-opening stage is from 16 to 21 February. Bud sprouting ratio is 78 percent and bud-growing ratio is 20 percent. Leaves mature from 15 March to 20 March, belonging to the medium sprouting and medium mature type. Autumn leaves harden in early October.

Average yield per ha: This variety has strong shooting ability with few side branches. From 1 m shoot 154 g leaves can be produced in spring and 132 g in autumn. There are 600 leaves/kg in spring and 250/kg in autumn.

Leaves are 36 percent of total weight of twigs, shoots, leaves and fruit. Annual leaf yield is 27 000 kg/ha.

Fruiting habits: Soroses are abundant, medium size, purplish-black.

Leaf chemical composition: Medium leaf quality. Leaf DM contains 21.5 percent of CP and 12 percent of SC.

Planting recommendation: The recommended training type is low or medium trunk.

Dao Zhen Sang

A native variety selected from Guizhou, diploid (2n = 2x = 28), belonging to the species of White mulberry (*M. alba* L.).

Morphology: Canopy is slightly spread. The twigs are medium sturdy, straight and long, greyish-white. Internodes are straight. Internodal distance is 3 cm. Lenticels are round and small, $21/cm^2$. Winter bud is triangular, light yellow in colour, slanting outwards. Accessory buds are few. Leaf is cordate, flatly spreading, light green in colour. Leaf tip is short caudate or sharp. Papillately dentate leaf margin. Emarginate leaf base. Leaf length is 18 cm, width 14 cm. Leaf surface smooth and slightly shrivelled, with lustre. Petiole is slim and short. Flowering type is staminate, bearing short and few inflorescences.

Climatic requirement: Bud sprouting in Jiangsu is from 3 to 9 April. Leaf-opening stage is from 12 to 17 April. Bud sprouting ratio is 75 percent and the bud-growing ratio is 28 percent. Leaves mature from 1 to 5 May, belonging to the early sprouting and early mature type. Autumn leaves harden in early September.

Average yield per ha: This variety has strong shooting ability with few side branches. From 1 m shoot 137 g of leaves can be produced in spring and 90 g in autumn. There are 968 leaves/kg in spring and 376/kg in autumn. Leaves are 50 percent of total weight of twigs, shoots, leaves and fruits. Annual leaf yield is 22 500 kg/ha.

Leaf chemical composition: Medium leaf quality. Leaf DM contains 22.01-24.58 percent of CP and 13.8-13.58 percent of SC.

Main diseases/pests and tolerance: This variety has medium resistance to Black wilt bacterial disease and resistance to Yellow-type mulberry dwarf. Its tolerance to cold is poor.

Planting recommendation: This variety is recommended for planting by roadsides and field boundaries, and to be trained into medium or high trunk.

Traditional Uses

Animal feeding: In China nearly 100 percent of the mulberry cultivated is used for silkworm feeding only. The litter, containing silkworm faeces and wasted leaf, is used to feed fish in the pond or as supplementary feed for cattle.

Silkworm feeding: Mulberry contains all the necessary nutrients for the growth and development of silkworm (*Bombyx mori*). For more than 5 000 years, mulberry leaf has served as the only feed for mulberry silkworm. Even today, this has not changed in China, although the utilization of an artificial diet has achieved encouraging results in the laboratory. Fresh mulberry leaf contains 70-80 percent water, 20-30 percent DM. In the DM, CP is about 25 percent, SC 25 percent and ash 10 percent. The leaf-silk inversion rate, or the feed efficiency of mulberry leaf for cocoon shell production, is the main index to evaluate leaf quality. In China, it takes 15 to 18 kg fresh leaves to produce 1 kg fresh cocoon at the farmer level.

Researchers investigated 251 silkworm varieties maintained at the Sericulture Research Institute of the Chinese Agricultural Academy and the results showed that the leaf-silk (DM) inversion rate is 10.11 percent in average.

Fruit: The fresh mulberry fruit (sorosis) has 85 percent water, 0.36 percent CP, 1.86 percent free acid, 9.19 percent invert sugar, 0.91 percent crude fibre and 0.66 percent ash. The fruit is also rich in carotene, Vitamin B_1, B_2, C, nicotinic acid, fatty oil, etc. The main sugar is glucose, and the main free acid is malic acid. The fresh fruit is traditionally consumed. In recent years, mulberry fruit juice has been commercially produced as a healthy beverage, and become very popular in China. Without adding preservatives, the juice of the mulberry fruit can be kept fresh under cold storage for three months, while the beverage can be kept fresh at ambient temperatures for 12 months.

Mulberry tea: The DM of mulberry leaf contains rich gamma-aminobutyric acid, which is as high as 266 mg/ 100 g DM in an average of 119 mulberry varieties investigated, about ten times higher than that of green tea. The main function of gamma-aminobutyric acid is to lower blood pressure and help nerve transmission. In China a certain amount of leaves are wasted by the end of the production cycle. At the end of the autumn crop, the unused leaf per hectare reaches 1500 kg or even more.

In order to exploit this valuable resource, the technique of preparing mulberry tea has been developed. The processing of mulberry tea is similar to that of green tea, which includes the procedures of leaf harvesting and washing, chopping, steaming, rubbing, baking, cooling, sieving and storage. Mulberry tea is drunk in the same way as green tea. Baked or fried mulberry tea powder is rich in protein and carbohydrate, and has a distinct fragant smell. It can be used as a food additive in the preparation of steam buns, bread, cakes and biscuits.

Edible fungi production: The dry stem of mulberry contains 50 percent cellulose, 20 percent semi-cellulose, 20 percent lignin and 5 percent CP. The proportion of carbon to nitrogen is 86:1, which is suitable for edible fungi production. By using the stem powder of mulberry as medium, mushrooms, Jew's ear (*Auricularia auricula judae*) and medicinal fungi (*Ganoderma lucidum*) have been popularly produced in China.

Medicinal uses: China has a long history in using mulberry as a traditional herb medicine. The medicinal function of the different parts of mulberry is as follows:

- *The leaf*: It is used against hyperlidaemia and diabetes.
- *The fruit*: It is good for the liver and kidneys, and is used against nephritis, constipation, etc.
- *The root lowers blood pressure*: The bark of the root is used against pneumonia and haemoptysis.

11

Mulberry Breeding, Cultivation and Utilisation

Mulberry has different varieties with male, female or androgynous flowers. Among the varieties with androgynous flowers, there are predominantly staminated, predominantly pistillated and even hermaphrodite types. Mulberry plants with anemophilous flowers cross each other easily and naturally.

The silkworm eats only mulberry leaves to make its cocoon, producing silk. Mulberry leaves are rich in protein and amino acids. It is known that there is high correlation between leaf protein level and production efficiency of cocoon shell, which means the cocoon shell weight to the total amount of mulberry leaves consumed by the silkworm. Therefore, an increase in the protein level of mulberry leaves may lead to improvements in cocoon productivity. Relationship between mulberry leaf protein content and the cocoon shell yield in spring and autumn

Mulberry was originally cultivated in Japan and other countries for sericulture. Recently, however, mulberry has been re-evaluated because of its functional characteristics and is being utilised for various purposes, such as the following:

Fruit: Mulberry fruit changes colour from green to purple black through red with maturity. Some varieties introduced from mid-Asia have white fruit. On average, the sugar content is about 12 percent, but in some varieties it is more than 20 percent. Mulberry fruit is consumed fresh, made into jam or liquor (mulberry wine). Very recently, it was found that mulberry fruit has an anti-oxidative property.

Medicinal uses: Mulberry has been used as a medicine from ancient times. The root bark in particular has been used as a herbal medicine to reduce high blood pressure. Mulberry leaf is rich in gamma-aminobutylic acid, effective against high blood pressure, and in alanine, effective against hangovers. Moreover, it has been found that deoxynojirimycin, which is said to have an effect in lowering the blood-sugar level closely related to diabetes, is abundant in mulberry leaf. That is why, today, mulberry tea is considered to be a health food.

Paper production: Mulberry grows more quickly than other woody plants and is said to be suitable for high biomass production. Mulberry branches are being used as raw material for paper production.

Mushroom production: Mulberry stem and stem powder are found to be a good source of media for mushroom production.

Mulberry for Poultry Production

If domestic and guinea fowls were to be raised in mulberry gardens, multiple useful effects could be expected such as weeding; pest control; extra fertiliser (from manure); high quality and natural egg and/or meat; and feed cost reduction.

Uchino et al. utilised guinea and domestic fowls (New Hampshire breed) for weed control. A small poultry house was built in the corner of a mulberry garden. Eleven domestic fowl were put in a mulberry garden of 0.025 ha and ten guinea fowl in one of 0.012 ha. Commercial feed was given, 140-150 g/dY for each domestic fowl and 70 g/day for each guinea fowl. The amount of weeds growing after seven months was measured.

Table 1 shows that there were no weeds in the guinea fowl plot, and only a few weeds in the domestic fowl plot, whereas a large number of weeds, such as Polygonum longisetum De Bruyn, Digitaria adscendens Henr. (large crab-grass), Malachium aquaticum Fries (Mater mouse-ear chickweed), Senecio vulgaris L. (common groundsel) and Veronica persica Poir. (Byzantine speedwell) grew in the control plot. These results indicate that raising guinea and domestic fowls in mulberry gardens controls weeds that would otherwise cause serious damage to mulberry growth. Similar results were obtained in the Oita Prefecture.

Table 1. Amount of weeds growing after seven months, fresh matter (g/m²) and percentage relative to control in parenthesis

Treatment	Amount of weeds	Weed species
Control	1 064 (100 %)	Polygonum longisetum, Digitaria adscendens, Malachium aquaticum, Senecio vulgaris, Veronica persica
Domestic fowl	11(11 %)	Senecio vulgaris, Erigeron philadelphicus
Guinea fowl	0(0 %)	

The daily excrement of domestic fowls is about 150 g, corresponding to 54.8 kg annually. It contains 1.6 percent of N, 1.7 percent P and 0.8 percent K. Uchino et al. measured non-organic nitrogen in three places in a mulberry garden. They found 12-14 times more in the front of the fowl house and two to five times more in between rows than in the control garden (Table 2). However, the quantity between stumps was almost the same as in the control garden. This study shows that both guinea and domestic fowls hang around their shelter. Spreading manure would be necessary to get a uniform fertilising effect.

Table 2. Non-organic soil nitrogen, mg/g of dry soil

Sampling site	Guinea fowl	Domestic fowl	Control
Front of fowl house	21.3	17.8	-
Between mulberry rows	7.3	3.7	1.5
Between mulberry stumps	1.3	1.9	-

However, a report from the Oita Prefecture indicated that the leaf yield in a garden that received only a quarter of N fertiliser and had domestic fowls, was the same as the control with 30 kg/10 area of N. Similar results were obtained in the Saitama Prefecture. It can therefore be concluded that domestic fowls in mulberry garden had a fertilising effect.

Uchino et al. investigated the pest control effect of guinea fowls by dissecting two birds and examining the insects inside their crops. They found that guinea fowls had eaten Japyx japonicum Enderlein, Aphodium rectus Motschusky, Arge similis Vollenhoven (azalea sawfly) and others. Moreover, daily observation confirmed that guinea fowls consumed Scotinophara lurida Burmeister (black bug), Spilarctia imparilis Butler mulberry tiger moth), Apriona japonica Tomson (mulberry borer), grasshopper, cicada, fry and earthworms. Apriona japonica, for example, is an insect that causes serious damage to mulberry growth in Japan.

Yearly egg production was 170 for a domestic fowl and 90 for a guinea fowl. Guinea fowl do not lay in winter. As shown in Table 3, guinea fowl eggs are smaller compared with those of domestic fowls, but they have a thicker eggshell and better preservation. Moreover, guinea fowl eggs are higher in Haugh unit and yolk colour. New Hampshire hens raised in a mulberry garden produced eggs with a greater proportion of yolk and higher Haugh unit and yolk colour, compared with eggs from commercial White Leghorn hens. These results demonstrate that eggs produced by guinea and domestic fowls raised in mulberry gardens are superior to those commercially available.

Table 3. Egg quality of guinea and domestic fowl raised in a mulberry garden

Species	Egg (g)	Yolk (%)	Yolk white	Shell (%)	Haugh unit	Yolk colour
Guinea fowl	39.2	31.4	0.60	16.6	83.1	13
Domestic fowl:						
New Hampshire	60.3	29.5	0.48	9.3	67.1	12
White Leghorn	59.0	27.5	0.44	9.5	61.4	10

In this experiment, after 23 weeks, domestic fowls, on average, reached 2.0 kg and guinea fowl 1.5 kg, their normal adult weight. Guinea fowl meat has an excellent taste good and is relished as a high-quality meat in

France and Italy, where it represents 30 percent of poultry meat products. On the contrary, guinea fowl meat is rarely eaten in Japan.

Effects of Mulberry Leaves

Several reports on the utilisation of mulberry leaves for poultry production have been published recently. When feeds containing 3 percent, 6 percent and 9 percent of mulberry leaf were fed to the domestic fowls (White Leghorn), egg quality (e.g. egg weight and egg production ratio) was almost the same as the control with commercially available feed. However, when the feed contained 15 percent of mulberry leaves, egg quality was significantly lowered. However, yolk was more yellow because of the mulberry leaves (Table 4).

Table 4. Egg yolk colour of domestic fowl with mulberry leaves in the feed

Feed	Yolk colour		
	Beginning	After 3 weeks	After 7 weeks
Control	8.0	8.5	9.2
15 percent mulberry leaves	7.8	9.6	10.1

Mulberry leaf in the feed increased vitamin K1 in the yolk (Table 5) but did not have an effect on the content of gamma-aminobutyric acid (GABA), which has a role in reducing human blood pressure (Table 5). There was no significant difference in the cholesterol content of egg yolk (Table 5). However, lipid peroxide content in the yolk significantly decreased after seven weeks of feeding mulberry leaves. It is known that lipid peroxide is closely related to active oxygen species, which have a role in causing diseases.

Table 5. Health-related compounds in the egg yolk of domestic fowl fed with mulberry leaves

Compound	Feed	
	Control	15% mulberry leaf
GABA (ppm)	1	1
Vitamin K1	1	54
Vitamin K2	22	29
Cholesterol (mg/100 mL):		
Before	15.1	15.1
After 3 weeks	13.9	15.6
After 7 weeks	12.9	14.3
Lipid peroxide (nmol/mL):		
Before	52.6	40.8
After 3 weeks	49.5	44.6
After 7 weeks	57.9	41.1

Table 6. Health-related compounds in blood serum of domestic fowl fed with mulberry leaves

Compound	Feed	
	Control	15% mulberry leaf
Cholesterol (mg/100 mL):		
Before	130.7	146.9
After 3 weeks	150.4	117.7
After 7 weeks	156.4	135.7
Lipid peroxide (nmol/mL):		
Before	11.6	12.1
After 3 weeks	12.8	13.2
After 7 weeks	13.9	11.3

Moreover, Sudo, Kuramoto and Iso, measured the effect of mulberry on the quantity of betacarotene contained in the egg yolk and blood serum of domestic fowls fed with mulberry leaves only. Beta-carotene content in mulberry leaves was 29 mg/100 g, which corresponds to the level in tea leaves. Betacarotene in the egg yolk and blood serum of domestic fowls fed with mulberry feed was 0.02 mg/100 g and 0.055 microg/100 g, respectively; meanwhile, in the birds receiving feed without mulberry (control) it was not detected. Betacarotene has an important antioxidant and anti-cancer effect and, when converted to retinol or vitamin A, has an anti-nyctalopia effect. Retinol concentration in the egg yolk and blood serum was the same in both treatments.

Table 7. Ammonium and hydrogen sulphate emitted from poultry manure

Feed	Odour compound (ppm)	
	Ammonium	Hydrogen sulphate
Control	2.67	9.14
10 percent mulberry leaf	0.29	10.89

Odours emitted from poultry manure, especially in peri-urban farms, have become a social problem in Japan. Ammonium and hydrogen sulphate are major odours in manure. Sudo, Kuramoto and Iso, tested whether mulberry leaves had an inhibitory effect on ammonium and hydrogen sulphate emissions from manure. Ammonium emission was significantly reduced by feeding mulberry, but hydrogen sulphate was not affected (Table 7). Suda obtained similar results on the inhibitory effect of mulberry leaves on ammonium emission. Thus, it can be concluded that mulberry leaves included in poultry feed have an odour reduction effect in manure.

Mulberry Germplasm in Brazil

The mulberry area in Brazil covers approximately 38 000 ha. The state of Paraná is the largest concentration of that area with 32 400 ha, followed by the western portion of the state of São Paulo, with 4 600 ha and by smaller areas in the states of Goiás, Mato Grosso do Sul, Minas Gerais and Santa Catarina. It must be highlighted that all production of these areas is for the silkworm, and no surplus is available for other uses. Plant management includes frequent pruning which prevents plants from blossoming and fruit development; plant sex is thus irrelevant.

According to Fonseca and Fonseca and Shammass, the first attempt to cultivate mulberry commercially occurred in state of Rio de Janeiro, as an initiative of the Emperor D. Pedro II, in the middle of nineteenth century. After a period of decadence, it was resumed only in 1923, in the eastern parts of the state of São Paulo State following Italian immigration. Later, eastern São Paulo showed a decline in sericulture and western São Paulo flourished, coinciding with Japanese immigration. Today, sericulture is concentrated in the northern part of Paraná State.

The causes that determined the movements of mulberry cultivation are not clear, but they must be related to various factors such as low income conditions prevailing among the rural population that sees sericulture as a secure source for income generation; the appearance of other economic alternatives; and soil exhaustion, production and profitability decline, resulting in sericulture opening space to other activities with more intensive in capital requirements.

In Brazil, all cultivated mulberry varieties appear to belong to M. alba. In São Paulo state, there is an active germoplasm bank (BAG) at the Estação Experimental de Zootecnia (Livestock Experimental Station) in Gália county; one collection at the University of São Paulo (UNESP) in Jaboticabal county; one in a spinning mill (under the trade name of Fiações de Seda Bratac) in

Bastos county; and one at the Agronomic Institute of Paraná (IAPAR), in Londrina. The Estação Experimental de Zootecnia has sericulture research as a mandate and its BAG includes three collections: variety collection (32 clones), the Instituto de Zootecnia collection (42 clones); and the Fukashi-Miura collection (14 clones).

Commercial companies have disseminated the Miura and Korin clones, because of the great supply of planting material. These clones together with the Calabresa variety, make up the large majority of the cultivated area. Genetically modified clones (IZ and FM) are very productive and with more nutritious leaves but, being destined completely to silkworm feeding, their expansion has been limited.

Cultivation Under Different Climatic Conditions

- *Temperature*. Optimal temperature required for mulberry is situated between 24-28°C; below 13°C growing stops and so do branching and bud formation. From August to May, mulberry vegetative growing is practically continuous.
- *Rainfall*. Ideal rainfall for mulberry This condition is met in São Paulo and Paraná states.
- *Relative humidity*. In general, the required air humidity for mulberry is between 65 and 80 percent. Higher precipitation and soil humidity are contributing factors towards improved leaf quality.
- *Photoperiod*. The major mulberry areas in Brazil are located between 20 and 30° south, and it seems that there are no photoperiodic limitations to plant growing.
- *Altitude*. In São Paulo State, in general, mulberry crops are located in altitudes varying from 300 to 700 m, far from the limit tolerated by mulberry.

— *Drought resistance.* No research data are available relating to drought resistance for varieties cultivated in Brazil. The seasonal supply of green leaves shown by some varieties even in drought periods was qualitatively observed for some clones, such as IZ 29/1.

Quality of Soil

Despite mulberry's need for adequate soil quality, it can grow in most of the soils available in São Paulo state. However, shallow soils and well as compacted, hard, impermeable and swampy soils should be avoided. Whenever possible, preferred soils are deep, fertile, well-drained, friable, porous, with adequate water retention capacity and a mild acidity (pH 6.2-6.8).

Fertilisation

Organic. In general, poultry manure is applied, with a minimum dose of 0.5 kg per plant, during the winter time, at a depth of 20 to 30 cm. Field experiments demonstrate higher productivity when mulch is applied to mulberry plants. When available, its utilisation is highly desirable, but some precautions are necessary. For example, when residues of different origins with very high C/N ratio are used, it is recommended that chemical N fertiliser be applied jointly in order to accelerate the decomposition of organic material and avoid competition for N in the soil.

Chemical fertilisers. According to the literature, there is no consensus on the ideal chemical fertiliser for mulberry. Among other factors, quantities will obviously depend on soil fertility. Soil analyses have been the most viable and practical way to evaluate soil fertility. Although there are no guidelines on the amount of fertiliser for mulberry according to soil fertility and economic factors, it is suggested that the table of Bulletin 100 from Instituto Agronômico de Campinas.

Table 8. Mineral content of organic materials

Material	C	N	P	K	Ca	Mg	S	Zn	Cu	Cd	Ni	Pb
Fresh bovine manure	100	5	2.6	6	2	1	1	33	6	-	2	2
Composted bovine manure	320	15	12	21	20	6	2	217	25	-	2	1
Poultry manure	140	14	8	7	23	5	2	138	14	2	2	17
Swine manure	60	7	2	5	12	3	-	242	264	-	2	3
Castor oil seed meal	450	45	7	11	18	5	-	128	73	-	-	-
Mucuna	60	3	0.6	3	2	0.4	-	6	3	-	-	-
Crotalaria juncea	70	2.8	0.4	3	2	0.4	-	2	1	-	-	-

According to Rubia Brasil Sobrinho and Azeredo, in cases of low contents in P and K and when all silkworm residues are applied to mulberry fields, the following application could be suggested (in kg/ha/year): 250 kg of N, 65 kg of P_2O_5, and 156 kg of K_2O. In manure utilisation, the data presented in Table 8 may help to determine nutrient application rates.

Mulberry Pests and Diseases

Pseudaulacaspis pentagona: This insect attacks mulberry branches and stems, introducing its mouth apparatus (stylus) into the sap plant conductors and sucking a great deal of the plant sap. These insects shield themselves under small structures and, when an attack is intense, these structures fully cover the branches and stems of the plant (Figure 1). The plant is quite weakened by the attack, showing cracks that allow the entrance of pathogenic micro-organisms. In boggy, shadowed and poorly aired soils, or in badly managed crops, this attack is more severe.

Figure 1. Mulberry branches infested by Pseudaulacaspis pentagona

Naupactus spp.: These beetles belong to the Curculionidae family. Adult phases of the insect feed on mulberry leaves, causing severe reduction in the leaf surface, and larvae phases feed on roots. As a result, plant vigour decays, once nutrient absorption collapses and pathogenic micro-organisms are eased through plant injuries. The Naupactus versatilis form has been more frequently detected in Paraná state. Its adult form presents a length ranging from 11 to 14 mm, with a metallic green colour, which becomes darker with age, due to the loss of scale-like structures.

Migdolus fryanus—Westwood-Coleoptera from the Cerambycidae family: This polyphagous beetle, biologically little known, has been causing economic damages in several crops, namely in sugarcane and, more recently, in mulberry. Adult males are generally black, dark brown or red brown, and their total length ranges from 12.1 to 37.0 mm. Adult females, on the other hand, are generally reddish brown or dark brownish, and their total length ranges from 19.5 to 35 mm. The male exemplars have membranous and functional wings, with antennas that reaches approximately half the body size. The females have reduced non-functional wings and much shorter antennae than the males.

Larvae cause the totality of the damage to the mulberry plants. Those larvae find their ideal source of nutrition in the root system, so destroying the roots. Insect dispersion happens in the larvae phases, because in this period an increased mobility in the soil environment is observed, establishing a net of channels that are likely to be used for adults in order to reach the soil surface during the mating period. Although plants are generally affected only partially, attacks can completely destroy the affected plants.

With regard to population control of the insecto, there are few solutions because of its biological and behavioural aspects. One obstacle is that larvae bury 3-4 m into the soil; eggs are laid at different depths. Although there was a generalised belief that attacks were more frequent in sandy soils, today studies on the subject have demonstrated that there is no linkage between types of soil and Migdolus attack.

"Cutting-prone" ants—Sauba ant.: Ant hills are true subterranean urban structures, endowed with a great number of individuals (more than ten million, in a adult ant hill), which can reach 7 m in depth, below the soil line. Quenquéns ants of the genus Acromyearmex are smaller than sauba ants and, in general, ant hills from this genus hardly reach depths lower than 50 cm. One ant hill communicates with another by ocelli (orifices on the

soil surface). In an ant hill of three or more years old, three types of ocelli are observed: one for transportation of land, one for ventilation, and one for food provision.

Nematode diseases—Meloidogyne spp.: These worm species attack mulberry producing root tumours (root knots), which make it hard for the plant to absorb nutrients and water causing severe damage to the crop (Figure 2).

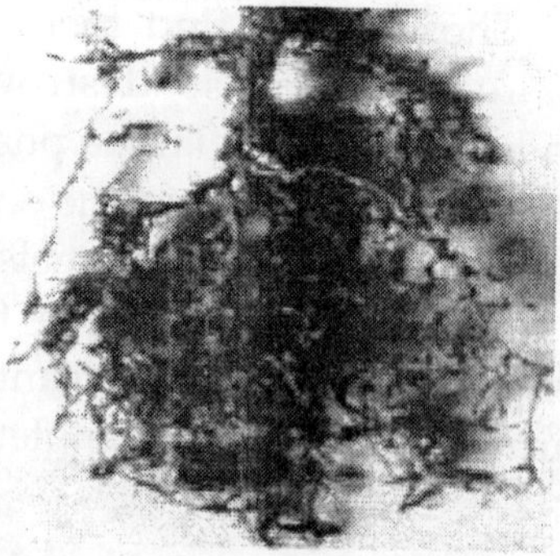

Figure 2. Mulberry root with root-knots

Plants attacked present reduction in size, with leaves appearing to lack water, becoming yellower and collapsing, reducing sensibly the production of leaves per area. Mulberry vitality deteriorates gradually and, in the later stages, the plant dies.

Main bacteria and fungus diseases

White root rot: This is caused by fungus from genus Rosellinia. This type of fungus lives as a saprophyte in decaying organic matter, mainly from plants left over the soil, and they can become pathogenic to cultivated plants. The diseased mulberry plants become weak, and the root rots. The hyphae emerging from the various fruiting bodies attack the smaller roots of the healthy plants, and spread to the main root system, causing white root rot.

Figure 3. Mulberry root with violet root rot

Violet root rot: This is caused by Helicobasidium mompa, Tanaka fungus. The First symptoms of this disease appear when mulberry does not sprout at the beginning of spring or when the leaves suddenly appear to wither in the middle of summe. The main disease characteristic is the existence of similar velvet veilings that colours the root bark dark-brown. Several reddish filamentous substances appear that colours all the root bark that are the hyphes from the pathogen.

Mulberry withering: This disease has been verified in mulberry cultivations of the state of São Paulo, mainly in the region of São José do Rio Preto. Its main characteristic is root rot. Several samples of attacked plants have been examined in research institutes and, to date, non-conclusive diagnoses are available. According to personal reports by scientific researchers, the following micro-organisms are detected in a sample from Guaraçaí, São Paulo: *Fusarium* sp., *Lassiodiplodia* sp., *Fusicoccum* sp., *Meloidogyne* sp. and *Pratylenchulus* sp.

Oidium: Phylactinia corylea Prest Karst. Oidium is a very common disease in mulberry plants and can cause severe damage to leaves, making them so they are unable to feed silkworm once the fungus has absorbed all the leaf nutrients. It frequently occurs in older leaves, which are covered by a greyish-white powder, made from pathogen structures. The disease is more frequent during the spring and summer, periods when favourable conditions for its development are present.

Rusty spot Cylindrosporium mori Targioni Tozzeti. In the beginning, the spots are small (less than 1 mm in diameter) but with the development of the process, they can reach more than 1 cm, and are brownish in colour with yellowish edges. The spots usuals to appear at the top, in the petiole and over the young branches. In severe attacks, mulberry leaves fall from the plant.

Other leaf spots caused by fungus. These attacks may be caused by fungus from the genera *Cercospora* sp., *Alternaria* sp., *Helminthosporium* sp. and *Colletotrichum* sp. Seasonally, they are more frequent during spring and summer, in periods of higher humidity and temperature.

Bacteria spots. Bacterium mori, Bacillus curbonianus (Macchiatti), *Bacterium moricolum.* Mulberry is susceptible to these kind of spots, at any age and in any aerial part of the plant. Seasonally, the disease is more frequently observed in summer (higher humidity and temperature). Initial symptoms appear as on the leaves small and sparsely distributed spots in necrosed areas, which coalesce and tear apart in a later phase, exuding yellowish pus on the lesion. The shootings can also die, causing oversprouting.

12

Classification of Livestock Production

Livestock production systems are considered to be a subset of farming systems. Livestock systems in which more than 90 percent of dry matter fed to animals comes from rangelands, pastures, annual forages and purchased feeds and less than 10 percent of the total value of production comes from non-livestock farming activities. Landless livestock production systems (LL).

Subset of the solely livestock production systems in which less than 10 percent of the dry matter fed to animals is farm-produced and in which annual average stocking rates are above ten livestock units (LU) per hectare of agricultural land. Grassland-based systems (LG). Subset of solely livestock production systems in which more than 10 percent of the dry matter fed to animals is farm-produced and in which annual average stocking rates are less than ten LU per hectare of agricultural land.

Livestock systems in which more than 10 percent of the dry matter fed to animals comes from crop by-products or stubble or more than 10 percent of the total value of production comes from non-livestock farming activities. Rain-fed mixed-farming systems (MR). A subset of the mixed systems in which more than 90 percent of the value of non-livestock farm production comes from rain-fed land use. Irrigated mixed-farming systems (MI). A subset of the mixed systems in which more than 10 percent of the value of non-livestock farm production comes from irrigated land use.

Landless Livestock Production Systems

The developed countries dominate the picture of landless intensive production with more than half of total meat production as shown in Figure 1. Asia is already contributing some 20 percent and eastern Europe 15 percent, with the latter recently in sharp decline. Landless monogastric production system (LLM). This system is defined by the use of monogastric species, mainly chickens and pigs, where feed is introduced from outside the farm, thus separating decisions concerning feed use from those of feed production, and particularly of manure utilisation on fields to produce feed and/or cash crops. This system is therefore open in terms of nutrient flow.

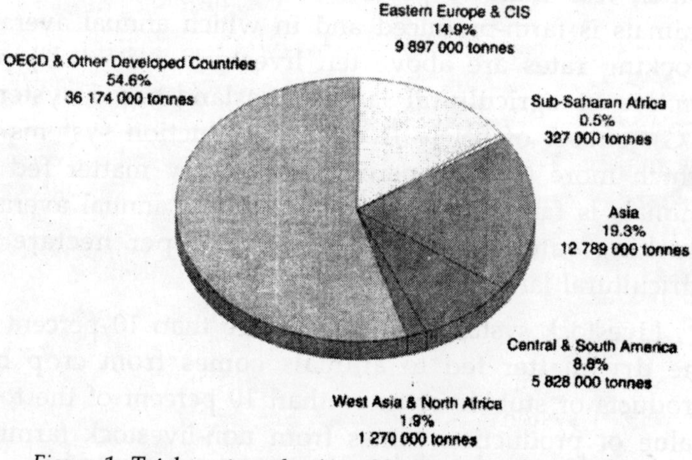

Figure 1. Total meat production of landless farming systems in different regions of the world.

Landless monogastric systems are found predominantly in OECD member countries with 52 percent of the total landless pork production and 58 percent of the landless poultry production globally. In the case of pig production, Asia is second, with 31 percent of the world total. For poultry, Central and South America follow, with 15 percent. To a large extent, this geographic distribution is determined by markets and consumption patterns in addition to levels of urbanisation. In Southeast and eastern Asia, this system is especially important. As much as 96 percent of the total pig-meat production in Asia occurs in China, Viet Nam and Indonesia.

China, Thailand and Malaysia produce 84 percent of poultry meat. This is associated with fast economic growth and urbanisation. The demand for monogastric meat is expected to grow from two- to fivefold between 1987 and 2006 from a base of 31 million tonnes, and a three- to tenfold increase is expected in the demand for eggs from 9 million tonnes. The prerequisites for development into large-scale vertically integrated production include the use of appropriate breeds and strains, feed quantity and quality, housing and disease control, as well as assured markets both at home and abroad.

Landless poultry and pig production systems account for the majority of the output in developed countries and their share is rapidly increasing in developing countries given their high supply elasticity in the short term. The system is typically competing with traditional land-based production systems for shares in the urban markets. It must be kept in mind that poultry and pork are close substitutes for beef and mutton, thus also interacting with the ruminant systems.

In a broader sense, the demand for cereals created by these systems is also competing for land resources with land-based ruminant systems. Landless ruminant production system (LLR). This production system is defined by the use of ruminant species, principally cattle, where feed is mainly introduced from outside the farm system. Landless ruminant production systems are highly concentrated in only a few regions of the world. In the case of cattle, they are almost exclusively found in eastern Europe and the CIS and in a few OECD member countries.

Landless sheep production systems are only found in western Asia and northern Africa. Typical cases are large-scale feedlots in the United States and in eastern Europe and the CIS, and veal production in parts of the KU. Intensive dairy operations in the same regions are more land-based because of the need to feed palatable fodder, which cannot be transported economically over long distances. In Asia, the intensive ruminant systems are typically found in buffalo and dairy cattle production units/colonies in India and Pakistan.

The LLR system is based almost exclusively on high-producing, specialised breeds and their crosses, which, nevertheless, have not been bred specifically for performance under "landless" conditions. With regard to milk production, the Holstein-Friesian breed is clearly the most important, and for beef production, English breeds predominate in the United States, while the large European dual-purpose breeds provide animals for fattening. The abundance of rangeland in the United States has led to the specialised production of calves from beef breeds for feedlot operations, while under European conditions these animals are a joint product together with milk, mainly from mixed systems.

The LLR system is highly capital-intensive, leading to substantial economies of scale. It is also feed-intensive and labour-extensive. Key efficiency parameters are daily weight gains and feed conversion, basically reflecting the efficient use of capital invested in infrastructure or in the form of lean animals and feeds. Weight gains are usually in the range of 1 to 1.5 kg per day, and feed conversion rates are about 8 to 10 kg of grains per kilogram of weight gain. This system is closely linked to land-based systems that normally provide the young stock for landless systems. This constitutes an important difference from landless monogastric systems, in which replacement stock is produced within the same vertically integrated system.

Driven by population growth, the landless ruminant system is expected to continue to grow slowly in North America and southern Asia. On the other hand, it is expected to decline in the EU as production becomes more extensive in response to policies reducing agricultural support. In eastern Europe and the CIS, its importance is also declining and ruminant production in that part of the world is shifting to land-based and smaller-scale systems.

A growing market for grain-fed beef exists in Japan and the newly industrialised countries of Asia. The growth rate of this market will depend mainly on the evolution of the international price of cereals and the increased per caput incomes. This market will in part be supplied domestically and through imports from the United States, Canada, Australia and possibly South America.

Grassland-based Systems

The importance of grassland-based systems in different world regions is shown in Figure 2. Central and South America and the developed countries dominate the picture in terms of meat production, together accounting for more than three-quarters of the world's production. Temperate zones and tropical highlands grassland-based system (LGT). In these areas, the grazing system is constrained by low temperatures. In the temperate zones, there are one or two months of mean temperatures, corrected to sea level, to below 5°C, whereas in the tropical highlands daily mean temperatures during the growing period are in the range of 5° to 20°C.

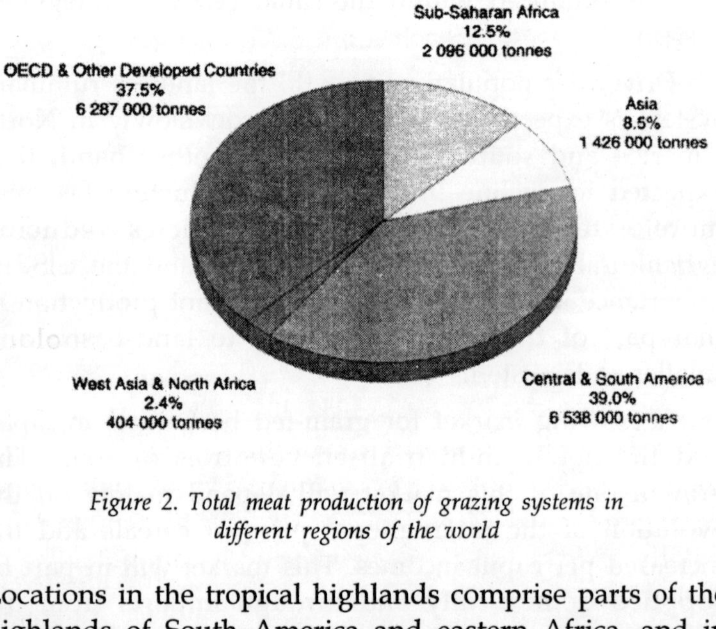

Figure 2. Total meat production of grazing systems in different regions of the world

Locations in the tropical highlands comprise parts of the highlands of South America and eastern Africa, and in temperate zones they include southern Australia, New Zealand and parts of the United States, China and Mongolia. Typical cases are Mongolia's steppe system, New Zealand's dairy and sheep enterprises, dairy systems close to Bogota, Colombia, and the South American camelid and sheep grazing systems in the altiplano of Peru and Bolivia. Extensive grazing systems are also found in parts of northwestern Pakistan involving sheep for mutton and wool and transhumant sheep on degraded high-altitude pasture in Nepal.

Further cases are reported for Chinese merino wool sheep on communal grazing in Jilin Province and sheep ranching on grass-clover pastures of intensive animal production enterprises in the state of Oregon in the United States. The regions in which the LGT system predominates have a combined human population of 190 million, which represents only 3.5 percent of the world total, and almost half of this population lives in Asia. In OECD member countries, far fewer people (14 million or 1.7 percent) use the LGT system, but they control more land and cattle per inhabitant than in the other regions.

Under the LGT system, product use varies widely, ranging from export-oriented New Zealand farmers, to South American farmers mainly producing for the domestic market, to Asian and African smallholders concerned with local markets and their own subsistence. Market forces and environmental concerns are curbing the potential for intensification of this system. As a result, their global market share is declining vis-à-vis that of other production systems. Since the LGT system is found mostly in marginal areas, its production potential in global terms is relatively low.

In developing countries it tends to form a subsistence basis for certain groups of the population and its future role is seen more in providing employment for these groups than in making a major contribution to output and economic development. In developed countries, frequently with production surpluses, the production from these systems is declining in relation to other values and uses assigned to these land resources, such as recreational value, value as a wildlife and biological diversity reserve and the contribution to water conservation.

Humid and subhumid tropics and subtropics grassland-based system (LGH). The LGH system is defined as a grazing system found in regions with more than 180 days of growing period. It tends to be concentrated more in the subhumid zone, particularly in regions where access to markets or, for agronomic reasons, crop production is limited. By definition, only very limited cropping is considered for subsistence. The LGH system is found mostly in the tropical and subtropical lowlands of South America: in the llanos of Colombia and Venezuela as well as in the cerrados of Brazil.

Dual-purpose milk-beef systems in the Mexican lowlands and estancias in Argentina are also typical cases of this system. In the African setting, many of the potentially suitable land resources are not used as a result of trypanosomiasis constraining livestock production. Outside Latin America, this system is important only in Australia because of its ample land resources in relation to its population. Worldwide, the LGH system comprises about 190 million head of cattle, an important proportion of which are of the zebu breeds.

In the subhumid and humid regions, cattle are clearly the dominant species, and in very high rainfall areas, such as the Amazon River delta and some parts of Queensland, Australia, buffaloes are also ranched. African hair sheep and dwarf goats are usually kept for local consumption only. In the subtropics, wool sheep are an important component of the system, for example, in Argentina, Uruguay, South Africa and Australia. The LGH system produces approximately 6 million tonnes of beef and veal and 11 million tonnes of cow milk worldwide. By far the most important geographic region is Central and South America. The system is predominantly market-oriented.

Globally, 6 percent of the world's population (330 million people) lives in areas where the LGH production system predominates. Its importance in terms of sustaining the livelihood of rural populations is expected to decline as interaction with crop cultivation turns it into a mixed system. In rain-forest regions, efforts are being made to incorporate perennial tree crops, frequently into silvo-pastoral systems.

In the savannahs, this system is being converted into a mixed-farming system by including annual crops, such as maize, soybeans and sorghum. Improvements in road infrastructure and new technologies to allow the establishment of pastures with commercially worthwhile nurse crops are making the ley farming systems, involving rotations of crops and pasture, a potentially attractive pathway into mixed-farming systems. Arid and semi-arid tropics and subtropics grassland-based system (LGA).

The LGA system is defined as a land-based system in tropical and subtropical regions with a growing period of less than 180 days, and where grazing ruminants is the dominant form of land use. This system is found under two contrasting socioeconomic frameworks: in sub-Saharan Africa and the Near East and North Africa regions, where it constitutes a traditional way of subsistence for an important part of the population, and in Australia, parts of western United States and southern Africa, where private enterprises utilise publicly or privately owned range resources for ranching purposes.

While in western Asia, northern Africa and sub-Saharan Africa, the LGA system is important for the livelihood of large sectors of the rural population, in developed countries it is extremely labour-extensive. In sub-Saharan Africa, agropastoralism is the most important interface between livestock production and other agricultural production. In other regions these systems are interrelated with other livestock production systems that have access to better-quality feed and are closer to markets. In low-income countries without an export market, incentives to produce quality beef are weak. This, in turn, limits the attractiveness of livestock production stratification.

Globally, new roles are emerging for these rangelands, besides that of producing ruminant animal products. In developing countries, the pressure to expand crop production is increasing population pressure on the remaining rangelands. In developed countries, the utilisation of these rangelands for animal production has often been subsidised through very low prices for grazing permits and public investments in irrigation.

Rain-Fed Mixed-Farming Systems

The geographic distribution of mixed-farming systems is depicted in Figure 3. Sub-Saharan Africa, West Asia and North Africa, and Central and South America are relatively unimportant in terms of meat production, whereas developed countries and Asia together contribute about 70 percent of the total meat production from mixed-farming systems. Temperate zones and tropical highlands rain-fed system (MRT). This system is defined as a combination of rain-fed crop and livestock farming in temperate or tropical highland areas, in which crops contribute at least 10 percent of the value of total farm output.

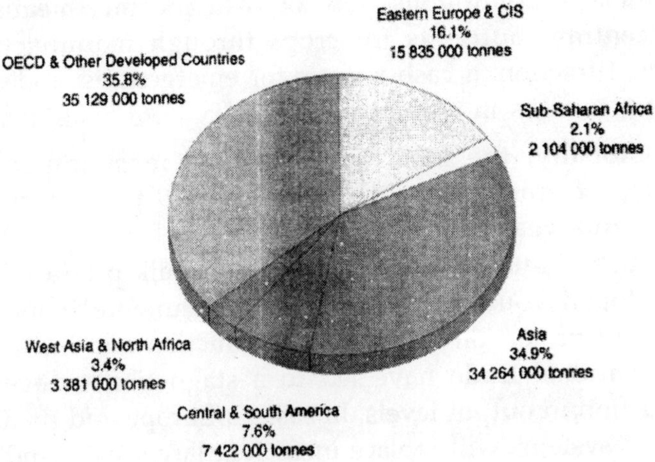

Figure 3. Total meat production of mixed-farming systems in different regions of the world

The MRT system is found in two contrasting agro-ecozones of the world: it is the dominant system in most of North America, Europe and northeastern Asia, basically covering large strips of land north of the 30° northern latitude parallel, and it is found in the tropical highlands of eastern Africa (Ethiopia, Kenya, Uganda, Burundi, Rwanda) and in the Andean region of Latin America (Ecuador, Mexico). The main common feature of these two regions is that low temperatures during all or part of the year limit and determine vegetation that is quite distinct from that found in tropical environments (e.g. C_3 versus C_4 grasses).

In the course of the development process, production technology in temperate locations evolved to higher specialisation, more use of external inputs and more open systems. This has resulted in increasing negative externalities of these systems for the environment. In most tropical MRT systems, production is less intensive, with livestock performing a series of functions in mixed systems: a continuous flow of cash income; means to concentrate nutrients for crops through manure; fuel; animal traction; a cash reserve for emergencies; and as a buffer to risks in crop production.

Globally, the MRT system is the most important source of animal products, providing 39 percent of the beef and veal production, 24 percent of the mutton production and 63 percent of the cow milk produced. In developed countries, growing environmental concern, reduced protection of domestic production and increased international trade have led to a stagnation or even a reduction in output levels. In eastern Europe and the CIS, mixed systems will replace inefficient large-scale landless systems as these economies open up to the markets.

In Asia, this system has the potential for increased production and better use of feeds for intensified ruminant production, associated with reallocation of land-use patterns at the farm level. In the highlands of sub-Saharan Africa and Latin America, production increases must come from a further intensification of crop-livestock land-use systems. The highlands should favour small-scale mechanisation because of the high population density, continuous cultivation and relatively heavy soils.

In the humid and subhumid regions of the tropics and subtropics, livestock production is based on mixed-farming systems. Given the range of socioeconomic conditions and soils and climates involved, this livestock system is very heterogeneous in its composition. It is found in all tropical regions of the world, mainly in developing countries. Parts of the southern United States are the only significantly developed regions included in this system. Typical cases are smallholder rice-buffalo systems in Southeast Asia or soybean-maize-pasture large-scale commercial operations in the Brazilian cerrados. This system includes regions with especially difficult climatic conditions for livestock (high temperatures and high humidity).

Adaptation of highly productive temperate breeds to these challenges has been notably poor. In many parts of Africa, trypanosomiasis constitutes an additional constraint to these systems. Particularly in African and Asian smallholder systems the local breeds are still widely used. In Latin America, Bos taurus cattle, sheep and goats were introduced some four centuries ago. Bos indicus cattle were introduced a few decades ago and have now replaced the earlier introduced cattle breeds in tropical areas. In the African and Asian MRH system, the multiple roles of livestock have prevailed, particularly animal traction and manure.

In Central and South America, this system caters to large domestic markets and, particularly in the case of Brazil, it is also linked to export markets. Under smallholder conditions, milk tends to be a more important output than meat. The MRH system applies to approximately 14 percent of the global population. This ratio is particularly high in sub-Saharan Africa, where 41 percent of the region's population is associated with the system, and in Central and South America, where it is 35 percent. The system is replacing grazing systems in Africa and Latin America. In Africa, the process is mainly driven by population growth, and, in Central and South America, by economic development and technological innovations.

The main challenge in sub-Saharan Africa is finding ways to increase productivity. It is generally acknowledged that the biological potentials of mixed systems will be the key to productivity increases, and the expectation is that purchased feed inputs will be replaced by nutrients cycled within the system. In Asia, increased crop production - and hence feed availability is an important way of intensifying and maximising animal productivity, however, the prospects for increasing cultivation area are limited.

In Latin America, low population density, a high degree of urbanisation and relatively high per caput incomes have resulted in farming systems that are generally more oriented towards livestock production. In the tropical rain-forest regions, high resource-consuming systems were established, in some cases driven by policies and in other cases by poverty. Many of the policies that promoted wasteful utilisation of these resources have been stopped in the process of structural adjustment. Arid and semi-arid tropics and subtropics rain-fed system (MRA).

The MRA system is a mixed-farming system in tropical and subtropical regions with a vegetation growth period of less than 180 days. The main restriction of this system is the low primary productivity of the land resulting from low rainfall. The more severe the constraint, the less important crops become in the system and the more livestock take over as a primary income and subsistence source. This system is important in the West Asia and North Africa region in parts of the Sahel (Burkina Faso, Nigeria), in large parts of India and in northeastern Thailand and eastern Indonesia, and less important in Central and South America.

Given the low intensity of the system and the multiple purposes of livestock, the introduction of improved breeds has been limited. Globally, 11 percent of cattle and 14 percent of sheep and goats are found in this system. Small ruminants are particularly important in

West Asia and North Africa under the MRA system. As is the case in other largely smallholder systems, livestock have a range of simultaneous roles in this system, including animal traction, production of manure and use as a cash reserve, in addition to the production of meat and milk. Fuelwood is often scarce as a result of deforestation and range degradation, leading to the ever-increasing role of animals as providers of manure for fuel, in addition to means of transport.

While this system supports larger populations than any other grazing system, only 10 percent of the world's population is related to this system. Fifty-one percent of the population involved is in Asia, mainly India, and 24 percent is in the West Asia and North Africa region. There is close interaction with the LGA system. With increasing population pressure, the LGA system tends to evolve into mixed systems, mainly MRA, because of the greater caloric efficiency of cropping vis-à-vis ruminant production when land becomes scarce.

The outlook for this system is relatively similar to that for the LGA system. The resource base puts a clear ceiling on agricultural intensification. Low and variable response to inputs makes their use financially risky. Population growth in this setting is contributing to the overexploitation of the natural resource base, as traditional property rights cannot cope with the growing demands. Alternative development strategies and the reduction of population pressure on the resource base are key elements for the sustainable development of these regions.

Irrigated Mixed-Farming Systems

Temperate zones and tropical highlands mixed system (MIT). This system belongs to the group of land-based mixed systems of temperate and tropical highland regions. It is found particularly in the Mediterranean region (Portugal, Italy, Greece, Albania, Bulgaria) and in the Far East (the Democratic People's Republic of Korea, the Republic of Korea, Japan and parts of China), where plant growth is limited both by low temperatures in the cold season and reduced moisture availability during the vegetation period. The system's importance in tropical highlands is negligible.

Meat, milk and wool, the main outputs of this system, are mainly produced for the market. Manure is an issue only where animals are stabled, at least for certain periods of the day or the year. About 10 percent of the global population lives in regions where this system is dominant. A large share belongs to developed countries with relatively high income levels and where agricultural trade is important. This system is clearly associated with very intensive agriculture in temperate regions with a high population density.

With the outcome of the General Agreement on Tariffs and Trade (GATT) negotiations, it can be expected that these systems will be less and less viable, having to compete with more efficient rain-fed systems producing the same commodities. The expansion of international trade, and particularly the incorporation of southern European countries into the KU, has led to an increase in intensive production systems for off-season vegetables and fruits on the best irrigated land. The integration with livestock has been reduced, with ruminant grazing systems declining in absolute terms and being concentrated on the marginal sites.

Humid and subhumid tropics and subtropics mixed system (MIH). This is a mixed system in tropical and subtropical regions with growing seasons of more than 180 days, and in which the irrigation of crops is significant. The MIH system is particularly important in Asia. High population densities require intensive crop production, and the irrigation of rice makes it possible to obtain more than two crops per year, even under conditions of very seasonal rainfall, substantially reducing yield variability as compared with the yield of upland rice or other rain-fed crops. In the past, animal production has been closely linked to the animal traction issue.

MIH systems throughout the world produce 13 million tonnes of pork (18 percent of global production), more than any other land-based tropical system. Among the tropical and subtropical systems, the MIH system is the one related to the largest population group, 990 million people, 97 percent of which are in Asia. Increasing labour productivity and relative affluence of farmers in this system are reflected in the more frequent use of tractors for cultivation. Manure is recycled on the fields.

Ducks do well in this environment, as do pigs and poultry, which thrive on abundant crop residues. In addition, since large ruminants, such as buffaloes and, to a lesser extent, cattle, have little association with draught and transport, their output will be increased milk and meat for the market. Animals and intensive crop production in this ecological zone are an illustration of a successful and sustainable agricultural production system. Competition for urban markets for livestock products is the main form of interaction with the landless monogastric system, both domestically and globally, through international trade.

Arid and semi-arid tropics and subtropics mixed system (MIA). This is a mixed system of arid and semiarid regions, in which irrigation makes year-round intensive crop production feasible. It is found in the Near East, South Asia, North Africa, western United States and Mexico. Typical cases are alfalfa/maize-based intensive dairy systems in California, Israel and Mexico; small-scale buffalo milk production in Pakistan; and animal traction-based cash-crop production in Egypt and Afghanistan.

Cattle and buffaloes for milk and animal traction are the main ruminant resources, although sheep and goats are important where marginal rangelands are available in addition to irrigated land. In the MIA system, pigs are kept only in the Far East; they are virtually non-existent in West Asia and North Africa, largely for cultural reasons (Islamic and Jewish religions). The main introduced breeds are dairy cattle to supply milk to large urban centres. Under good management conditions, intensive dairy schemes have been quite successful in hot but dry environments. Some of the world's highest lactation yields are achieved in the MIA system in Israel and California.

The traditional smallholder MIA system in Asia relies heavily on buffaloes for milk production. In the traditional MIA system, irrigated crop production is the main source of income, with livestock playing a very secondary role. This is generally reflected in rather extensive management of livestock enterprises. The MIA system is predominant in regions that are home to over 750 million people, two-thirds of them in Asia and one-third in West Asia and North Africa.

The main interaction with other systems occurs through the international market, particularly for milk and dairy products. The MIA system makes an important contribution to food availability and employment in semi-arid and arid regions. Irrigation allows increased fodder production as a by-product or part of crop rotation, which reduces the feed deficit. The improved feed base and utilisation promotes intensification and commercialisation of livestock production, especially in areas with good market access.

13

Fodder Trees and Shrubs

Vast areas of the earth's surface sustain a vegetative cover whose shrubs and small trees provide feed for domestic livestock and wildlife. By various adaptations, shrubs survive and even thrive under environmental conditions so difficult they preclude highly productive grasses and forbs. Le Houérou pointed out that nearly one third of the world's land surface is natural grazing land and to varying degrees the shrub component of these grazing lands is an important source of animal feed.

In West Africa, Boudet pointed out the valuable role of shrubs in supplementing the low nutritional value of standing grasses during the dry season. Increasing pressures on the land could be ameliorated by planting *Acacia albida* and bamboo to serve also as windbreaks. In North America, shrubs of the Great Basin desert rangelands provide 50 to 70% of the diet of sheep and 40% of cattle that graze these lands during the winter. At the Great Basin Experimental Range, Holmgren and Hutchins found this salt desert shrub range type to be an adequate "holding area" for maintenance of breeding or gestating livestock that utilised the living twigs of dormant browse plants in the winter.

Forti reported that introduction of shrubs to the northwestern Negev could create a new ecosystem to provide high palatability browse but also interact with certain low-growing introduced species to increase their productivity. He also pointed out an often-recognised fact that shrubs could withstand and survive under intense grazing. Gasto reported using *Atriple repanda* in an artificial forage plant community structure which optimised space by stratifying levels of plant overstory and thus increase the potential productivity. A report by the US National Academy of Sciences declares that tropical legumes are important resources for the future.

Information assembled on *Cassia sturtii, Desmanthus virgatus, Desmodium* species, Honey locust and *Leucaena* describes the opportunities for use and some management practices presently employed. Names of references who can be reached for specific information are listed. In addition, the report describes how *Acacia* and *Prosopis* species in several world locations are managed and used for multiple purposes. Le Houérou summarised data from various world locations indicating a high dependence of grazing animals on shrubs to satisfy their protein requirements, especially during seasonal dry periods. Without such shrub grazing in dry seasons to complement the forage available during moist seasons the entire cycle of livestock use would be in jeopardy.

Firewood

Cutting or grubbing out shrubs to be burned as a fuel for cooking and heating constitutes an important use. More important, this use seriously depletes the vegetation resource. Desperation or ignorance may lead to the removal of all shrub and tree species regardless of their values if left in place. In his research on the ecology of desert shrub rangelands in Iraq, Thalen pointed out the destructive nature of fuel cutting on shrublands bordering settlements. He estimated 25 kg per day to be a family requirement for cooking and heating in cold seasons.

Van Der Veen estimates that 40,000,000 *Haloxylon* shrubs, equal to a thousand hectares of dense shrub vegetation, are burnt yearly in Syria to prepare tea! Woody fuel gathering and transport back to villages is a common sight in developing countries where petroleum-based fuel sources are unavailable or too expensive. Unfortunately, as pressure mounts for an increase in production of charcoal greater emphasis shifts from woodlots to the use of trees and shrubs of rangeland areas. In some areas where the shrub resource is already depleted by removal for fuel or by overgrazing such additional pressure is disastrous.

Changing long-standing cutting practices such as restricting cutting to top growth so that the plants can regrow may be difficult. Combined and unrestricted use of shrubs for both grazing and cutting for fuel is a destructive practice. Thalen pointed out that grazing alone will not destroy shrubs because livestock will move on before all vegetation has disappeared leaving the woody stems to regrow. Fuel gathering, although more restrictive in area, destroys the remaining crown at the surface level.

A more expensive alternative is to plant desirable species that may satisfy fuel as well as other needs such as browse or soil fertility maintenance. Where dense shrub growth exists removal of the less desirable species for making charcoal would at the same time release good browse and forage grass species from competition and stimulate their growth. Energy from woody biomass is being investigated as an alternative fuel for industrial use through pyrolysis or bioconversion to a liquid or gas fuel. Van Epps et al have found that selected shrub populations growing in favourable sites of the Great Basin can produce as much as 458,000,000 Kcal/ha of energy.

Shrubs are generally equal in heat value to low quality bituminous coal. Other indices of biomass yield from arid lands range from 10,000 kg/ha/yr for *Atriplex canescens* to 28,160 kg/ha for *Prosopis chilensis* in the Imperial Valley of California under irrigation. However, under less favourable conditions an annual yield of 4,000 kg/ha was considered a reasonable production.

Maintaining Soil Fertility

A recent publication by ICRAF, the International Council for Research in Agroforestry, focuses on the role of trees and shrubs in maintaining soil fertility in forest/ agricultural ecosystems. Legume trees and shrubs offer a low cost alternative to maintenance of soil fertility in developing countries and areas where the expense of artificial fertilisation would not be economically feasible. According to ICRAF the focus of soils research should be on studies in marginal lands, rehabilitation and protection of disturbed ecosystems, improving soil and water conservation, investigating the efficiency of traditional agroforestry practices, evaluating the efficiency of N_2 fixation and ways to exploit the process, study "nutrient pumping" by deep rooted perennials, determining optimum fallow period, and developing ways to use woody perennials to stabilise agricultural systems.

Charley reviewed various aspects of how favourable soil fertility resulted from the presence of shrubs. He pointed out that soil nutrient information on arid ecosystems was both geographically and biologically diverse. Most of the soil nutrient capital is near the surface and not in the sub-soil except in relation to rooting pattern. Le Houérou gave several examples using agroforestry techinques to enhance pasture and crop production. Felker and Bandurski characterised an ideal food crop tree as one that would concurrently enhance soil nitrogen levels.

Shrubs and trees have been used in various cropping practices where they were deemed critical to the maintenance of soil nitrogen and other elements essential to plant growth. This application is limited in value to specific areas but generally advocates of high intensity agriculture have ignored any potentials for contribution of legume trees and shrubs to soil fertility. Desert fertility is intimately tied to the shrub species that make up its vegetative cover.

Garcia-Moya and McKell showed that nitrogen was concentrated in the soil under the canopy of shrubs and in shrub tissues, thus constituting "islands of fertility" in the desert ecosystem. West and Skujins reviewed sources of nitrogen in desert ecosystems and the processes by which it is cycled. Shrubs play a vital role in this cycling process. Destruction of shrubs, especially those that promote symbiotic N-fixation can cause a significant reduction in ecosystem productivity. Numerous reports have been made of the favourable soil productivity impact of *Prosopis cineraria* in agropastoral systems in Pakistan.

Shankar et al., reported that range grass productivity was found to be nearly doubled under a *P. cineraria* canopy relative to other legume trees. In addition to an increase in soil nitrogen, other soil characteristics such as pH, organic matter, P_2O_5 are also more favourable under a tree of shrub cover. The Dehesa live oak woodland of the Spanish Estremadura is managed in a mixed farming system of winter cereals and cultivated fallow, which are followed by several years of pasture in which diminutive legumes add modest amounts of nitrogen.

The natural populations of trees and shrubs serve as a stabilising agent in the system although 2,000 plus years of taking off more soil nutrients in crops than are cycled back home left this system producing at a bare fraction of the original potential. Increased productivity as well as a shift in species composition from *Bromus mollis* to *Bromus rigidus* is attributed to the presence of blue oak *(Quercus Douglasii)* in the annual range type of California whether elimination of blue oak as a range improvement practice would ultimately reduce the level of forage production is not known but may be expected based on observations from other ecosystems.

In the areas of the Sahel where annual rainfall permits marginal cropping, the increased nutritional status of the soil under canopies of *Acacia albida* makes possible a two and one half times better production. One of the most striking examples of potential soil nutrition enrichment by a shrub/tree species is that of *Leucaena leucocephala.* Reports from field trials in Hawaii indicate a benefit of 500–600 kg/ha/yr of N if the foliage harvested at 3-month intervals from this plant were mulched into the soil. It may be possible to develop drought resistant strains of this species as indicated by survival of one-metre tall plants introduced at the Kiboko Experiment Station in Kenya by an FAO project on genetic resources. However, soil nitrogen contributions under arid conditions by *Leucaena* are not known.

Wildlife Habitat

Habitat that provided suitable cover, food and behavioural activities is critical to maintaining wildlife populations. Many countries in Africa depend heavily on a tourist industry that is based on the viewing of unique wildlife species. Big game species and upland birds are constrained by the size and quality of habitat that is under pressure from agriculture and community encroachment. Areas degraded by over use may need to be restored. Some animals have very specific habitat requirements involving certain plants and plant densities without which their populations would decline.

A recent compilation of habitat requirements in the arid and semi-arid United States to guide habitat rehabilitation indicated that some animals are restricted to few vegetation types while others use a broad range of types. Competition between wildlife and livestock for forage and browse has caused serious problems in many areas of Africa, especially in areas close to National Game Parks. Vegetation management such as clearing shrubs and trees in a corridor around game parks to deter migration or better control of fire to increase shrub density may be a means of reducing the areas of conflict.

However it is less likely that increasing shrub density would deter big game species because shrubs are generally important to their cover and escape requirements. On the other hand, continued herding livestock into bomas for protection is often a necessity in many areas where carnivorous wildlife occur. Opportunities for complementary browsing should not be overlooked. For example, eland thrive on browse and, whether domesticated or not, this species could exercise some control over woody vegetation not preferred by other animals.

Jensen et al., have shown that grazing by sheep or cattle for a short period in the early growing season on areas reserved for wildlife can provide additional grazing units without harming browse-dominated vegetation. Studies of this type are needed when conflict for livestock use and wildlife habit-use occurs.

Control of Desert Encroachment

The International Conference on Desertisation in 1977 brought to the fore some of the critical locations where desert encroachment threatens villages, roads, structures or agricultural areas with drifting sand and soil particles. Equally important was the discussion of physical processes and inappropriate resource management practices that accelerate the processes of desertification to marginal lands in many world locations. Shrubs can be important in the restoration of soil stability and vegetation cover of lands destabilised by improper use.

Shrubs interfere with wind currents near the soil surface thereby increasing turbulence and cause wind born soil particles to be deposited. In addition, shrubs stabilise the surface soil and retard aeolian soil movement. Plantings of saltbush and other drought and salinity-tolerant shrubs effectively stabilised an extensive area of wind erosion in Iran. Substantial plantings of *Acacia albida* and other species in the Sahel hold promise for future land stability. The selection criteria and establishment methods described by Van Epps and McKell may provide some guidance to land stabilisation practices.

Of all the multiple uses of browse species in arid lands, provision of livestock feed appears to have the highest priority because of the animal products that are subsequently produced. Wise management of livestock grazing use can facilitate additional uses.

Medicinal Uses

Numerous medicinal uses are made of arid land trees and shrubs. Some properties are of documented values while others may be mythical and legendary. McKell et al., described some medicinal and industrial uses of range plants in a symposium in Monterrey, Mexico. Some of the uses they discussed varied from gum arabic obtained from *Acacia senegal* and used in the cosmetics and pharmaceutical industry as adhesives and carriers, to latex obtained from Guayule *(Parthenium argentium)*.

Interest is now high in the potential for obtaining a quality lubricating oil from Jojoba *(Simmondsia chinensis)* which could replace scarce sperm oil from whales. Gooding and Northington include many examples of new products from arid land plant resources in their symposium proceedings. Another review of new plant products from arid lands was published by the US National Science Foundation and included some interesting products from shrubs such as berries from *Sheperdia argentea,* latex from Guayule *(Parthenium argentatum.* pine nuts from *Pinus edulis,* herb leaves and grain seeds from *Atriplex hortensis.*

Industrial Use

Although not generally considered as a suitable building material for large structures, shrubs and small trees are used considerably by people in rural areas for houses, shelters, animal pens, and fences. Without shrubs mud-plastered walls of many structures would lack the strength provided by woody stems woven among the supporting poles. Many useful implements and household tools are made from sticks and stems from trees and shrubs. Bark and fibres taken from woody plants also find use in construction as a framework and matrix for baskets, shelters and chinking material in walls made from poles or boards.

Windbreaks

Large shrubs and small trees offer protection from the erosive force of wind by serving as windbreaks. Even in areas where trees are less adapted the wind-breaking force of shrubs is important to prevent the destructive action of windblown sand particles. Planting of windbreaks on the margins of fields, roads, livestock holding areas and settlements opens up several opportunities of multiple uses such as browse, food, and fuel. In such areas there is a good likelihood of favourable plant growth as a result of increased soil fertility associated with the deposit of fine soil particles as well as water harvested from adjacent areas where plant growth is restricted by traffic or use.

Landscaping

Landscaping with native shrubs can provide an improvement in natural beauty and only requires minimal maintenance. Although landscaping gets little if any attention from people under poverty conditions, the presence of multiple use shrubs around a dwelling or livestock pens can shade the structures from the direct rays of the sun and also provide food and some protection. *Opuntia* species serve well in these conditions.

Nutritional Values of Fodder Trees

Various parts of browse species have been chemically analysed at different development stages. A major contribution to understanding browse chemical composition was published thirty three years ago. Subsequently, browse data of a wide variety have been published. Additionally, many reports of chemical constituents have been included in descriptions of browse species from various regions. Given the impressive list of publications dealing with chemical composition of browse plant tissues, an important goal is to determine the significance of such values in light of large differences in free-choice preference of animals for individual browse species and plant parts.

One important problem is to determine the nutritional relevance for each livestock species of such measurements as total nitrogen, nitrogen free extract, ether extract and others. Extensive feeding and digestion trials make it possible through regression equations to determine the nutritional significance of certain types of analytical data when unique sources of feed such as browse are being considered. Clearly, much work needs to be done to determine the true nutritional value of browse species that are used at various seasons by particular animals.

For example, sagebrush *(Artemisia tridentata)* has a relatively high energy content but not all of the indicated energy value is available to an animal because a portion of the energy-yielding base is from volatile oils which may not be easily metabolised by the animals but more importantly, their presence reduces plant palatability. For various reasons, animals eat some shrubs in preference to others, as has been shown by Dougall and Bogdan and Dougall et al. Spines, rough surfaces, stout stems, salt content, volatile oils or various other characteristics may be responsible for such palatability preferences as exercised by various types of animals.

Hanks et al., have shown that sagebrush plants with a high phenolic content were often the least browsed by deer in winter. Other substances and/or characteristics need to be better understood if valid selections are to be made of populations within species for multiplication to be used in seeding or planting. Intensely grazed plants, however are a good indication of high palatibility – unless animal numbers are much in excess of the carrying capacity of the range and forced grazing is occurring. Species or plants within species that can sustain intense browsing should be emphasised in grazing land plantings.

Shrubs provide their greatest benefits to grazing animals during periods of nutritional stress when the nutritional value of dormant grasses and forbs is low or when there is little forage except shrubs available because of drought. The value of browse shrubs was central in Bulletin 10, "The Use and Misuse of Trees and Shrubs as Fodder". Authors of that publication recognised the widespread value of shrubs as a stability factor in grazing land productivity but suggested that areas of high potential or where bush encroachment had occurred could be made more productive by substituting a grass forage.

Millions of hectares, however, are stable only if they support a combination of shrubs, small trees, and forb-grass understory and should be utilised only under a programme of ecological management. Some of the gaps in knowledge mentioned in *Bulletin 10* have been filled, but the hope for improved range conditions which would include shrubs has not materialised due to drought and increased animal grazing pressure. This challenge is still with us but to meet it will require palatability evaluations and management opportunities. For example, Fort concluded that sheep preferred groups of species rather than single species. Further, he showed that the 14 shrub species he studied had excellent ability to recover after severe browsing.

Progress is being made as evidenced by plantings of large blocks of land to fodder shrubs in North Africa, Syria, Iran, and other locations. Many shrub species have been recognised for their values as a green feed during periods of forage deficiency as well as other uses. Le Houérou reviewed potentials of fodder trees and shrub species from a world-wide forestry perspective.

Improvement programmes for multiple use grazing lands should include plants of the highest possible biological potential. Information on shrub biology is deficient in many areas, however, and difficult to obtain from numerous worldwide sources. Computer-based data retrieval systems often fail to yield desired information because of inappropriate or inadequate designation of key words related to the nature of the research. Authors preparing key-words for their manuscripts should take care to properly identify the major points of their work and include appropriate designations of environment and use such as "arid land", and "grazing use".

One desired outcome for scientists attending this symposium is their greater awareness of information sources and access to additional data about the biology of high potential shrubs. Field observations have provided many data about animal use as it occurs under certain ecological conditions. From this information we have learned to predict a potential range of adaptation and possibilities for expanded use. Thus, we can describe what species exist in an ecological niche with some degree of success and to a lesser degree what we must do to put the biological potential of specific browse species to good use(s).

Because of our persistently inadequate knowledge on the biology of shrubs, however, we know very little as to how certain management practices function or how to make them more effective. This lack of background information prompted the publication of a fact-filled applied bulletin entitled, "Underexploited tropical plants with promising economic value" which included data on several shrubs that have multiple use possibilities. Valuable information on shrub biology was assembled in the *Proceedings* of the 1971 International Shrub Symposium and in the comprehensive treatment of shrub radio-ecology and ecophysiology by Wallace and Romney.

The extensive reports of research from the International Biological Program (IBP) Desert Biome represent another source of basic information. Several synthesis volumes from the IBP Desert Biome project are available or in press, and provide new information on deserts and their ecological functions. The general insufficiency of data for many geographical areas and subject matter topics represents a challenge to the present symposium. Dr. Le Houérou has identified a list of knowledge gaps that should receive a high priority for investigation. Gaining knowledge in these priority areas should further assist the development of multiple use management practices.

A useful conceptual approach for researchers seeking ways to develop browse shrubs for multiple uses is to keep in sight the relationship of the specific research to the biology of the species and how derived information can be put to use. Because of the opportunities, if not the necessity, for multiple use, the implications for related applications need to be borne in mind. For example, this interrelationship is the guiding force in a new research project at Utah State University which concentrates on "ecosystem reconstruction" following drastic disturbance. In this project soil physical-chemical processes are seen as being of direct concern to the establishment of shrubs and grasses individually and in early plant community distribution patterns.

Plant survival is conditioned by a host of interacting factors ranging from soil processes of moisture supply, mycorrizal inoculation, and nutrient availability to aerial factors of temperature, insect parasites, and seedling competition. Because they must plan for post disturbance land uses of watershed, livestock grazing, and wildlife habitat after achieving a stabilised hydrologic system the research team is not only given rigour but also meaning to their pursuit of the project. The intent of all browse research should include: to make the browse system work better, produce more effectively and be a stabilising influence on the ecosystem.

Few arid and semi-arid lands used for grazing and other purposes are currently producing at optimum levels. With improved management these lands could be more productive and stable. However increased understanding of available management practices and the biology of plants and animals is needed. Undoubtedly social changes must also be initiated but these must be worked out in conformity with local customs and values. Five major questions must be addressed to ensure the optimum use of shrub-dominated grazing lands. These problems are:

1) How much intensity of use or kinds of use are possible within the capacity of plants to maintain productivity?
2) How can undesirable plants be kept from invading or increasing in areas where desired species are under intense use?
3) What measures are available to control poisonous plants?

4) How can a desired balance of species be maintained to provide optimum productivity and ecosystem stability?

4) What are the best methods of establishing or replacing desired species to meet multiple use requirements?

Capacity to sustain use depends to a great degree on the inherent biological constitution of the plants involved. Critical periods of phenological/physiological development should be understood and grazing avoided at those times. Replenishment of carbohydrate reserves must be considered in determining the intensity and timing of defoliation. George showed that continued removal of new growth of snowberry (*Symphoricarpos oriophilus*) depleted carbon reserves to the point that plants died even though a portion of stored carbohydrate was still present, but apparently not available for translocation from storage tissues.

Various experiments in clipping browse at various times of the year have shown that regrowth can be stimulated as long as growing points are not destroyed. Willard found that the highest number of new shoots and the most favorable shoot survival was associated with a simulated management/utilisation schedule resembling rest-rotation grazing. Removal of the bark (Phloem) of cork oak *(Quercus suber)* is an example of timing the harvest of a product that cannot be repeated too frequently or both a low quality product and damage to the plant would occur.

Lopping off shrub or tree stems for emergency feed in Australia has its limits. The frequency of severe pruning of oak trees in Spain must allow time for recovery in order to assume production of acorn crops. Proper use factors were developed for certain grasses, forbs and shrubs as a guide to grazing management and to estimating the number of animals that could be allowed to graze a rangeland unit. Such proper use factors were often developed from observation rather than experimentation but the principle remains the same. Standards must be developed and used in relation to any management plan that seeks to optimise a multiple use plan.

Control of undesirable plants is a necessity where differential use within a plant community places an extra burden on the desired species. Uncontrolled plant competition plus grazing pressure will result in an increase in the less palatable species or those that are not grazed. Control measures may be drastic such as selective herbicides or mechanical means as reviewed by Vallentine, or more subtle methods such as changing the season of use or livestock. For example West et al., found that in rangelands grazed by sheep in the fall the average longevities of major grasses were enhanced, whereas the longevity of three-tip sagebrush *(Artemisia tripartita)* was reduced. Where shrub species of low value tend to dominate their abundance can be reduced by concentrating goat use on target areas as reported by Provenza to have occurred in southern Utah in controlling blackbrush *(Coleogyne ramosissima)*.

Control of poisonous plants can be a serious problem because of the lack of selectivity in available chemical and mechanical methods. Control by hand labour is a good possibility in areas of low cost labour but the gained security value for grazing use must be worth the control costs. Selectivity by animal preference is a possibility if based upon careful observation. For example, *Solanum incanum* is known as a poisonous plant to cattle in East Africa but it is palatable to goats and is thus amenable to control through management if conditions warrant.

Sheep grazing desert shrublands in the Great Basin of the western USA must be carefully herded and managed so as to avoid their over-consumption of the poisonous weed *Halogeton glomeratus,* which contains oxalic acid. Low amounts of ingested *Halogeton* can be tolerated if taken in with other forage species. Chemical control of the weed is impractical because of its widespread occurrence but close relationship to *Atriplex* species which are also members of the family Chenopodiaceae.

Shrub-dominated grazing lands can easily shift in botanical composition to a less-productive state if proper management of some degree is not applied. Differential grazing preferences will reduce the vigour and/or numbers of palatable species and thus allow less palatable plants to exert undue competition. In a grazing study extending over 30 years on the desirable species, Holmgren and Hutchins found that plant composition shifted in response to season of use, intensity of use and type of livestock use. Grazing by sheep tended to encourage grass dominance whereas cattle grazing encouraged shrub dominance.

Late winter grazing resulted in the loss of valued spring forage species. A highly palatable species, *Artemisia nova* which was used under heavy winter grazing, has been replaced by *Chrysothamnus stenophyllus,* a species of low palatability. Even within a species, the different strains-or biotypes that are more palatable will tend to decrease in proportion. Parker theorised that old shrub plants in a rangeland site were the least palatable ones within the population and represented the end product of palatability selection. He further recommended a fast-chaining as a moderate range improvement practice to uproot the old brittle plants but leave the younger, lower, pliable plants that have had less selective pressure for palatability.

Grazing management practices that reduce the selectivity of animals for a given species while leaving others free to compete with the grazed plants can help maintain a favourable species balance. Corbett allows livestock to graze for a short duration and does not return them to the same area until it has recovered. Some adaptation of a high intensity-low frequency or rest scheme may help to minimise grazing induced plant succession.

Multiple Uses of Plants

Many areas where shrubs are needed for various uses have been depleted of their woody cover. In spite of the aggressive nature of shrubs and their high adaptability to arid conditions they cannot survive excessive use. Reestablishment of an effective shrub population is thus necessary even though seedling establishment is difficult because of drought, disease and destruction by insects and small mammals. Considerable information is needed about shrub seedling biology to sustain a revegetation effort. The woody plant seed manual provides data on seed characteristics, areas of adaptation and germination requirements on many useful shrubs and trees of North America.

Such data would be helpful if available for other continents. Addition of new shrubs to a plant community can be obtained by natural regeneration only under rare circumstances when ecological conditions are favourable. At such times the competition between perennial and ephemeral species is great and unless the favourable moisture and temperature regime is sufficient the perennial species perish. Two alternatives to waiting for natural regeneration are seeding and transplanting. Seeding may be accomplished by broadcasting seed by hand or from a vehicle but is of minimal effectiveness and wastes seeds unless they are covered by soil.

Seeding with a drill implement is more effective than broadcasting but is limited to relatively level topography. Plummer et al., have tested and recommended 10 principles for restoring big game range. These principles included developing a rational seeding plan for suitable sites of adequate precipitation and for using a sufficient quality and quantity of adapted species in a season of highest suitability. Even with drill seeding, the chances for success can be low in areas of low and erratic rainfall. Bleak et al., reported numerous failures in seeding grasses in the salt desert shrub vegetation type of the Great Basin and recommend native shrubs as an alternative to grasses.

Better techniques for drill seeding are needed to make this relatively low-cost method effective for arid and semiarid lands. Transplanting of container-grown seedlings or cuttings is a high labour alternative to drill seeding. Because older seedlings are used the hazards normally eliminating tender young seedlings are not as critical. Even so, care must be taken in propagating container-grown plants and in planting them sufficiently deep in moist soil to assure survival. Planting large blocks of fodder shrubs requires an organised programme of seed preparation, planting in containers, nursery or greenhouse growth of seedlings and field planting.

Cooperation of a forestry nursery unit to provide field-ready seedlings as reported by Draz in Syria can help to increase chances for success. Involvement of local people in planting the seedlings gives them a sense of project participation and concern for the success of the project. Limited information or seed sources may result in a decision to plant a single species but this should be avoided if possible. A mixture of ecologically adapted species is generally more desirable than a single species —especially if that species is from a single collection or location. Protection from use for a suitable period of time is essential to allow plants to gain an adequate root system and plant vigour.

Subsequently, the intensity of use should not exceed the capacity of the planted species to sustain use. The attitudes of local people need to be considered in planning the development programme. This caution would hardly seem necessary but without a management plan that prevents overuse, shrub and tree plantings cannot survive as has been seen in a number of locations. Involvement of local farmers and livestock growers has the advantage of training them in the management of shrubs and gives them an appreciation for protecting seedlings until they are ready for use.

Bibliography

Adeola, O., "Nutrient management procedures to enhance environmental conditions: an introduction", *J. Anim. Sci.*, 1999.

Boschini F.C., "Establishment and management of mulberry for intensive forage production", *Proceedings of electronic Conference*, 2000.

Coffey, M.T., "A swine integrator's perspective on nutrient management procedures", *J. Anim. Sci.*, 1999.

Goering, J.K. & Van Soest, P.J., "Forage fiber analyses (apparatus, reagents, procedures, and some applications)", *Agric. Handbook* No. 379, Washington, DC, USDA, 1970.

Harris, W.D. & Popat, P., "Determination of the phosphorus content of lipids", *Amer. Oil. Chem. Soc. J.*, 1954.

Jayal, M.M. & Kehar, N.D., "A study on the nutritive value of mulberry (Morus indica) tree leaves", *Indian J. Dairy Sci.*, 1962.

Kitahara, N., Shibata S. & Nishida T., "Management and utilization of mulberry for forage in Japan 1. Productivity of mulberry-pasture association system and nutritive value of mulberry", *Proceedings of electronic conference*, 2000.

Liu, J.X., Jun Yao, Yan, Yu, B. J. Q, Shi1, Z.Q. & Wang, X.Q., "Nutritional value of mulberry leaves and their use as supplement to growing sheep fed ammoniated rice straw", *Proceedings of electronic conference*, 2000.

Lohan, O.P., "Cell wall constituents and in vitro DM digestibility of some fodder trees in Himachal Pradesh", *Forage Res.*, 1980.

Majumdar, B.N., Momin, S.A. & Kehar, N.D., "Studies on tree leaves as cattle fodder", 1. Chemical composition as affected by the stage of growth, *Indian J. Vet. Sci.,* 1967.

Narayana, H. & Setty, S.V.S., "Studies on the incorporation of mulberry leaves (Morus indica) in layers' mash", *Indian Vet. J.,* 1977.

Sánchez, M.D., "World distribution and utilization of mulberry, potential for animal feeding", *Proceedings of electronic conference,* 2000.

Schmidek, A.,Takahashi, R., Nunes de Medeiros, A. & de Resende K. T., "Bromatological composition and degradation rate of mulberry in goats", *Proceedings of electronic conference,* 2000.

Singh, B. & Makkar, H. P. S., "The potential of mulberry foliage as feed supplement in India", *Proceedings of electronic conference,* 2000.

Subba Rao, A., Amrith Kumar, M.N. & Sampath, S.R., "Studies on mulberry (Morus alba) leaf stalk palatability, chemical composition and nutritive value", *Indian J. Vet. Sci.,* 1971.

Ye, Z., "Factors influencing mulberry leaf yield", *Proceedings of electronic conference,* 2000.

Index